D0384531

NO LONGER PROPERTY OF
SEATTLE PUBLIC LIBRARY

SNOOZE

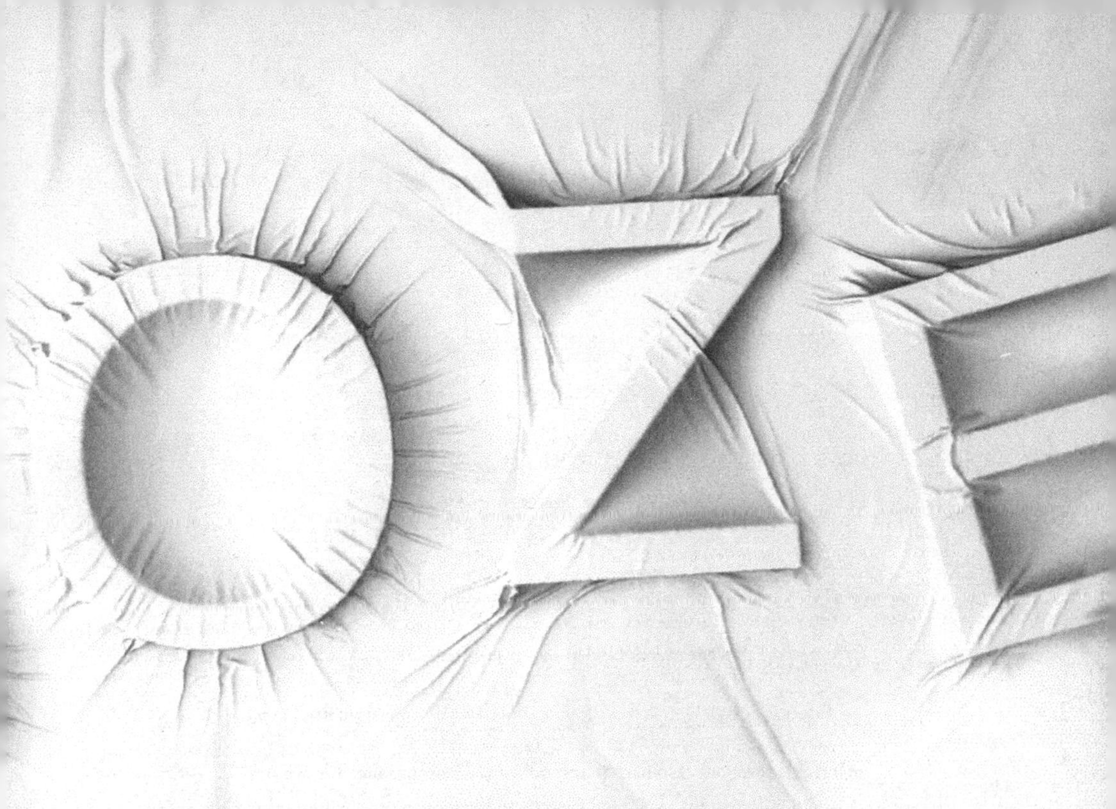

The Lost Art of Sleep

Michael McGirr

PEGASUS BOOKS
NEW YORK LONDON

SNOOZE

Pegasus Books Ltd
148 West 37th Street, 13th Floor
New York, NY 10018

First Pegasus Books hardcover edition June 2017

Interior design by Sabrina Plomiatallo-González, Pegasus Books

ISBN: 978-1-68177-419-0

10 9 8 7 6 5 4 3 2 1

Printed in the United States of America
Distributed by W. W. Norton & Company, Inc.

For Benedict, Jacob and Clare

And for Tony Flynn

"O sleep! it is a gentle thing,
Beloved from pole to pole!"

—*The Rime of the Ancient Mariner,* Samuel Taylor Coleridge

"I only know that while I sleep I have no fear, nor hope,
nor trouble, nor glory. God bless the inventor of sleep."

—*Don Quixote,* Miguel de Cervantes

"I wish I could write a chapter upon sleep . . . and yet,
as fine as it is, I would undertake to write a dozen chapters
upon button-holes, both quicker and with more fame,
than a single chapter upon this.

—*The Life and Opinions of Tristrum Shandy*, Laurence Sterne

INTRODUCTION

8PM

[1969]

Spare a thought for Rip Van Winkle, the man who slept through the American Revolution. His story is one of a number of fairy tales in which sleep is a spooky kind of experience. In this world, sleep can start with a curse from an old fairy, end with a kiss from a young prince, and last anywhere up to one hundred years. Such fables have always made curious bedtime reading—the purpose of which is surely to encourage children to surrender to the world on the other side of their closed eyes, not to be frightened of it.

Like many people, I first met Rip Van Winkle in my pajamas. It was 1969, and our bedtime was 8:00 PM. Mum had read in *Women's Weekly*, or somewhere like that, that the astronauts who went to the moon had begun life by always being in bed by 8:00 PM. We were going to follow their example, at least as far as sleep was concerned, so there we were, tucked in bed when Rip Van Winkle arrived. He came after our prayers, traveling in a large volume of children's stories that had the cover falling off. In this volume of wonders, people did all sorts of strange things that never happened anywhere near our rather staid family. They slept in forests and castles and glades (whatever they were). There was even an old woman who lived in a shoe. She had so many children that she could sleep anywhere, I suppose.

In an attempt to avoid his nagging wife, Rip Van Winkle went to sleep just before the American Revolution and then slept for twenty years. He woke up in a United States of America that had successfully untethered itself from the British monarchy and found that he himself had also been freed from his domestic obligations, as his wife had died while he was asleep. Two decades of sleep is a mediocre effort by the standards of Sleeping Beauty and various others. But, still—in an age well before we'd coined that turn of phrase FOMO (fear of missing out)—I remember being somewhat disturbed by old Rip's story. Sleep had deprived him of a ringside seat in one of the greatest moments in history. Was this supposed to be a fairy tale or a horror story?

This is a book about sleep. In the history of human civilization, sleep is the unrivaled hero. It is the wellspring of creativity. In sleep, we are most ourselves because we have to surrender our egos. It is the space in which so much happens, mainly because, while we are asleep, we can't squeeze any extra appointments or make any extra phone calls or look at one more thing on the Internet. Sleep is the daily visit that people who live in crowded diaries make to a wide-open space.

Sleep has piqued my interest because it has been a point of struggle for me throughout my life. I have lived with a few significant sleep disorders, starting with a diagnosis of sleep apnea twenty years ago when I was working as a Jesuit priest. I had spent most of my life being tired when my wife and I—already parents of a one-year-old—had twins. It was then that I learned what sleep deprivation truly was. But this is not a guidebook. I am not a clinician of any sort. I won't be telling you when to go to bed, how to prepare for bed, what to eat before bed or when to turn off your TV, phone, iPad, iPod, computer, or any other gadget. (The irony of modern living is the image of someone sitting up in bed looking up sleep techniques in the search engine on their computer. It's as silly as driving to the gym. Worse. It's like drinking to achieve sobriety.) If you do want help with sleep, there are plenty of good books available, full of sound advice.

This is not one of those books. My only clinical advice is pretty basic: If you are struggling with insomnia or nightmares or a sleep disorder, try to read a bit and talk to someone before automatically reaching for pills as a solution. Be wary of giving sleep problems a medical name before they really need them; insomnia is a pathology whereas poor sleep is just an inconvenience. The pair can look similar. It requires wisdom to know the difference; our problems often grow into the names we choose to give them.

This is a book about more than sleep. It is an exploration of the role of sleep in our history and culture. There is some information about my own experience, but I also reach more broadly to look at some of the most famous wakers and sleepers in history. It is also a book about the opposite of sleep—the phenomenon of *not* sleeping and what this kind of exhaustion means for society. The whole world is like an overtired baby. It screams at us incessantly when what it most wants is a decent sleep.

There's an old saying that says insomnia is nothing to lose sleep over. That is a simplistic statement, but it does contain a grain of truth. So instead, in the wee hours of the morning, when sleep eludes you, you now have this book.

8:48PM

[1876]

In the end, it only took one man to change the lightbulb.

Thomas Edison patented over a thousand bright ideas in his lifetime, although his name is remembered for maybe half a dozen of them. But the ones with which he is commonly associated are household items that changed the way people lived.

By the age of fifteen, Edison—who had a paltry formal education and a noted inability to sit still, but a habit of reading books compulsively—left his home in Port Huron, Michigan, to become a telegraph operator. The telegraph had at last solved the problem that had cost the runner his life after the battle of Marathon: no longer did messages need to be delivered in person. And Edison loved it. As a teenage telegraph worker, he won standing among his colleagues for the speed of his work and for his willingness to work nights, meaning they didn't have to. While others drank, Edison studied. He read everything about electricity and warmed to the work of Michael Faraday—a man, like himself, who had started in obscurity. Faraday, author of *Experimental Researches in Electricity*, had also quit school early; he then used the spare time left over from menial jobs to inch his way through the *Encyclopedia Britannica*. In 1831, Faraday created the first electric generator. He turned electricity from a mysterious, even spiritual, abstraction into something that could be manufactured and sold, a commodity that only needed the right master to put it to work. Edison would become one of the first of those masters and the most entrepreneurial.

It was never going to be enough for young Tom (Al to his friends, after his middle name, Alva) to master Morse code and pass on news of other people. He wanted to be the news himself. He studied the technology of the telegraph and saw where it could be improved, coming up with a quadruplex—a means whereby a single wire could be used to send not one but four messages at a time, two in each direction. This had obvious appeal to the telegraph companies, who now only had to roll out a quarter of the lines they did before. One company, Western Union, had contributed money to the development of Edison's invention and was wrong-footed when, at the last minute, Edison sold the idea to a rival, Atlantic and Pacific, for $30,000—a princely sum in 1875. The boss of Western Union, William Orton, said famously that Edison had a vacuum where his conscience ought to be.

In 1876, Edison—now married to Mary Stilwell, eight years his junior— bought an estate at Menlo Park, New Jersey, twenty miles from New York. The estate became known as his "invention factory," and Edison himself was labeled in the media as "the wizard of Menlo Park." Edison liked that. His pragmatism always came wrapped in mystique. He knew that genius was simply a matter of putting nuts and bolts together in the right order, but he was only too happy when his nuts and bolts roused a kind of spiritual awe in his admirers.

The reporters kept coming to Menlo Park, and the wizard kept sending them away with stories of electric pens and megaphones and other wonders put together by his remarkable team of nutters and bolters. In 1876, he came up with the idea that turned Alexander Graham Bell's awkward fledgling telephone into something even more useful. The new device could not only convey sound, but it could also record it. In 1877—to the dismay of Bell, who was honing in on the same conclusion—Edison patented the phonograph (the word literally means "sound writer").

The public was dumbfounded that a machine could speak. The human voice had never been stored before; it had always been a thing of the spirit. Now it had been made physical, engraved on a foil drum. Cartoonists had a field day: one of them showed a wife using the new toy to stir her husband in the middle of the night with cries of "help" and "murder." Edison's assault on sleep was already underway. When he demonstrated the new toy at the White House, President Rutherford B. Hayes stayed up all night, playing with it into the wee hours.

Edison knew that he was tapping into a rich vein of mysticism; the human voice could now be held for all eternity. Future generations could listen to people long after they were dead. Later, when Edison was working on early versions of the movie projector, he said he was trying to do for the eye what the phonograph had done for the ear. He was loosening Father Time's grip on his sickle.

Edison's first wife, Mary, endured a lonely marriage. Given Edison's work schedule and penchant for late nights in the lab, she dined on her own most nights and nearly always went to bed alone. She died of nervous exhaustion—deemed "congestion of the brain"—in 1884 at the still-tender age of twenty-nine. Although her death took place at two in the morning, Edison had to be fetched home from work to take his place at her bedside. He was, to be fair, distressed by the loss and broke into tears. But his tears may have been salted by guilt. As Mary's nerves had worsened over the previous five years, Edison had invested himself, day and night, into his dream of taming the dark.

The year after Mary died, Edison proposed to Mina Miller. Like Mary, Mina was sixteen when Edison suggested the merger. He declared his intentions by tapping "Will you marry me?" in Morse code on Mina's hand as they rode side by side in a carriage, and Mina understood enough of his world to be able to tap back her reply. At least she knew what she was in for.

Thomas Edison had little personal use for most of his inventions. He taught Bell's gurgling telephone to enunciate clearly and thus turned it into a viable economic proposition. But he would never come to the phone himself, least of all when his wife was trying to find out what time he'd be home. He spent years on a futile project to develop a type of rubber that could be grown in North America to make pneumatic tires (Henry Ford and Harvey Firestone were friends), but he never drove a car himself. He loved the phonograph, but he was so deaf that he could only experience the machine by biting into the wooden table on which it sat and feeling its vibrations. He pioneered cinema technology but didn't go to the movies; he preferred to sit up late and play Parcheesi instead—or, of course, to work.

Electric light was different. That was something he could use. For Edison, there were never enough hours in the day. So he made more.

One of the wonders of Edison's long life is that he died in his bed. He had spent so little of his eighty-five years in it that the odds were that he'd die someplace else. When he did get into bed, he often neglected to take off his boots.

Edison's approach to sleep is the stuff of legend. From an early age, he developed the habit of catnapping—snoozing for a few minutes at any odd time during the day. His signature posture was sitting at a desk with his hand balled into his cheek, supporting his head while he dozed. One of the most famous photos of Edison shows him fast asleep, flat out on a laboratory bench with his boots tied and his hand in his watch pocket—ready, it would seem, to check the time the moment he stirred. His hand is under his chin as though, even asleep, he is still working on a problem. At Menlo Park, he had a hidden cupboard under the stairs where he could retreat for a nap. His ability to sleep anywhere, from the side of the road to lying on the cowcatcher on the front of a train, may have been helped by his inability to hear.

Edison didn't like other people sleeping either. A contemporary biographer, George Bryan, relates the story of a job applicant telling Edison that he had terrible problems with insomnia. Edison employed him on the spot and set him to work, asking him to keep watch on an experiment "day and night." After only sixty hours on the job, his new employee dozed off, and Edison was furious that the man had misled him about his need to sleep. In later years, Edison went on camping jaunts with Henry Ford and Harvey Firestone. On one such jaunt, they found themselves in Vermont and decided to drop in on President Calvin Coolidge. Mrs. Coolidge told the visitors that the president liked a nap after lunch and was usually tucked up in bed early at night. Once again, Edison was not impressed. He is reported to have told Mrs Coolidge to tell her husband that "lack of sleep never hurt anybody." In his seventh decade, Edison wrote to a fan that the secret of his success was that he had worked eighteen hours a day for the previous forty-five years. "The body is only a piece of machinery," he said.

Edison's obsession with work was not easy for the colleagues he dragged out of bed at all hours because he was determined to move a project to the next stage of development. At a quarter to ten, his inner circle of men would be far from thinking of going home. After all, it was only two and a quarter hours to lunch—always served at midnight—and usually accompanied by beer, cigars, and male antics. In the 1880s, after Edison had brought electric light to New York and moved his office there, he often had lunch in Delmonico's restaurant on the corner of Fifth Avenue and Fourteenth Street—right on the dot of midnight.

Despite what he said, it's hard to avoid the conclusion that Edison was habitually exhausted. Certainly, he was always on the cusp of sleep. His most famous adage is both a recipe for exhaustion and an expression of it. In 1903, he told a chemist who had just joined his team that he didn't believe in luck. He said that he never quit anything until he got what he was after. He quoted to the chemist the wisdom of his favorite authority—himself. The adage was "Genius is 1 percent inspiration and 99 percent perspiration."

Edison was undoubtedly a genius, but his celebrated slogan is by no means an accurate representation of the way he divided his own energies. He doesn't cut any piece of the pie for bravado, self-confidence, bluster, and an ability to give his undivided attention to a dozen things at once—all core elements of the Edison persona. The belief that genius is 99 percent perspiration encapsulates an approach to invention that became the cornerstone of the 20th century's idea of research and development. Edison always found the one thing that would work by eliminating all the others that wouldn't. He was never threatened by the prospect of looking for a needle in haystack; it was easier than imagining a needle that didn't exist yet. Edison didn't work in the abstract; at Menlo Park, there was a storage room filled with every possible chemical compound, half-finished circuit, and piece of wire that ever came into the place—any of which could come in handy someday. Edison kept notes on all his inconclusive experiments, which he declined to think of as failures, knowing that they could contain the answer to a question that had not even been asked up to that point. Nothing was thrown away. If you worked through the rubble long enough, you would eventually find what you were after. You only had to stumble into it. Even a tired person can do that. It's possible to work through a haystack, straw by straw, while you are all but asleep on your feet and far beyond any capacity for thought. Edison expounded a way for people to be exhausted and still look as though they were coming up with fresh ideas. In essence, he invented the modern career.

And the wizard indeed had to sift through a mountain of rubble to change his lightbulb. At the time, a lot of money had been invested in gas lighting, which released toxins and blackened walls. Electricity had been used to provide what was known as "arc lighting," a form of illumination that required the combustion of carbon rods and thus let off fumes and heat. It produced a powerful light that couldn't be dimmed or regulated; it was only suitable for outdoor use or perhaps use in large enclosed spaces, such as factories.

Domestic light needed to be less obtrusive. Edison wondered if the practice of incandescence, or heating something until it glowed, could be improved upon. The idea had already occurred to Joseph Swan, a chemist from the north of England, who had been fiddling around with incandescent bulbs for almost twenty years. (Edison had no problem taking advantage of minds other than his own. In the mid-1880s, he would promise $50,000 to the newly arrived 29-year-old Croatian electrical genius Nikola Tesla, if Tesla could iron out certain problems in the generation and delivery of electricity. Tesla worked through the night for a year and achieved remarkable results. When he then asked Edison for the money owed to him, Edison told him he had to get used to the American sense of humor. Tesla didn't laugh. He ended up working for Edison's archrival, George Westinghouse. Swan was gobsmacked by the speed with which Edison developed not only a bulb but a whole system of fittings, sockets, meters, and lines for electricity distribution. The point was not to have electric light in a laboratory but to have it in every home.

The year 1879 in Edison's life bears comparison with the last year of Mozart's. The chap was busy. In September 1878, Edison began experimenting with a long list of substances that might provide a suitable filament to do the glowing inside a bulb. He and his associates raided the attic in their search and, in order to coax more money out of investors, prematurely announced that they had succeeded. They tried cardboard, cedar, celluloid, coconut hair, coconut shell, cork, cotton soaked in various substances, and more. Then, in January 1879, Edison started fiddling with something new. He tried making a filament glow inside a vacuum, inside a bulb. In March, to keep the press quivering, he announced that he was about to halve the price of gas lighting; he was still miles from his mark. He covered a vast stretch of ground in the next six months. But his resolve never flickered. He knew that invention was only a matter of discovery.

On the night of October 22, 1879, he got a bulb to burn for over twelve hours; within days, he had one that burned for one hundred hours. On December 28, Joel Cook, a reporter for the *Times* of London, which took the story to the other side of the Atlantic, wrote with amazement, "I ate my supper and wrote the draught of this telegram by Edison's light Saturday night." It would not be long before eating by candlelight became something special. By Christmas, the world was beating a path to the magical kingdom of Menlo Park. Over the next few years, Edison would move heaven, earth, and New York City to create a commercial network for electric light. Now the whole world could stay up late.

Like Macbeth, Edison had murdered sleep. He was part of a gang of assassins who have been at work since light divided night from day.

9:00PM

[700BC]

Of course, we sleep because we are creatures. But trying to understand what sleep is all about has been one of the engines that has shaped our culture.

One of the stories that gave birth to western culture is Homer's *Odyssey*, a work that continues to provide fodder for comic strips and popular culture almost three thousand years after it started to take shape—somewhat longer than the average life expectancy of books these days. Though Homer's identity is lost in the shades of time (it's possible that Homer is an amalgam of different people who contributed to an oral tradition), legend has long held that the visionary Homer who sang *The Odyssey* into life was blind. That may be why *The Odyssey* whispers in the dark.

At its root, *The Odyssey* is a book about getting home to bed. The hero, Odysseus, leaves the rocky island of Ithaca, where he is king, and goes off to fight the Trojan war. He makes an appearance in Homer's prequel, *The Iliad*, where he has the clever idea of using a wooden horse to smuggle soldiers into the besieged city of Troy. Once the war is won, Odysseus has to get back home—and this is the plot of *The Odyssey.*

It takes him quite a while to find his way home; by the time he gets back to Ithaca, he has been away for twenty years. His queen, Penelope, has been waiting patiently all that time, beset by a pack of sleazy suitors who have wanted her to agree that her husband is dead and team up with one of them, a prospect she doesn't relish. Like Odysseus, Penelope is in search of rest. For three of the years Odysseus is away, she uses her insomnia to keep the suitors at bay, telling them that she can't possibly marry until she has completed the shroud for her father-in-law, Laertes, which she works on every day, crafting a "great and growing web." At night, she stays awake, undoing "by the light of torches" the work she has done on the shroud that day, in order to give herself an excuse for the work continuing the next day. (Perhaps even more effective than telling would-be suitors she was shampooing her hair.)

Penelope is more faithful than Odysseus, who, along the way, has been held for seven years by the nymph Calypso, who has forced him to have sex with her. The divine Circe also requires sex from Odysseus against his will and only lets him go on condition he visits the underworld, where he encounters his dead mother; it must have been a blow to Circe's self-esteem that he accepted such terms. There is no account of Penelope being told these parts of the big adventure, and whoever Homer really was, it's not hard to see a male hand at work in these fantasies. Being forced into sex by a goddess is at the risky end of the spectrum of excuses for being home late.

Odysseus survives endless experiences, some of which—like the story of how he tricked the one-eyed Cyclops, Polyphemus—are familiar even to children whose classical education depends on McDonald's placemats. Eventually the king does turn up in Ithaca again, dressed in rags and looking like a beggar. Only his dog, Argos, recognizes him. Twenty years is long enough for a human but even longer for a dog. Argos used to be a fine hunting hound but is now a bag of bones. When he sees his master back home, he thumps his tale in anticipation. Sadly, the excitement is too much and the old pooch drops dead.

Odysseus will eventually clean out the suitors and be reunited with Penelope just in the nick of time. But the way the story is told, the point to which Odysseus returns is not his island, nor his throne nor his banquet hall nor his wealth nor his servants nor even his wife, although Penelope and Odysseus do have nice welcome-home sex. The point toward which Odysseus has been traveling through storm, disaster, bloodshed, and confusion has been his bed. When Penelope requires positive proof that this stranger is her husband, she asks, within Odysseus's earshot, for their bed to be shifted. Only Odysseus would know that the request is impossible. He gets angry and says that of course their bed can't be moved. The reason for this is that he himself carved it in situ from the stump of an old olive tree whose trunk had taken centuries to reach the width of a pillar, and he built the walls of their chamber around it. Clearly, Odysseus had never wanted his wife to waste time mucking round with the furniture.

The bed has roots deep in the earth. This is an evocative image, one which gives *bedrock* a literal meaning. Odysseus's bed is part and parcel of the rocky soil on which he was born. The entire world of his travels is anchored to it.

A quick flick through a paperback version of *The Odyssey* soon shows how carefully the theme of sleep is incorporated into the bones and sinews of the poem.

Odysseus's friend and guide among the gods, Athena, is the bringer of sleep. Odysseus gets into bother for sleeping in the wrong place, such as when he nods off within site of Ithaca itself but before his work is done—that is, without having made landfall. His sailors take the opportunity to look at his loot and untie the bag he has been given, which contains the four winds. The ship is blown back to square one and Odysseus must endure yet more suffering. But when the right moment comes and Odysseus does make final landfall, he immediately falls into a deep, delicious sleep on the sand, undisturbed by the memory of what he has been through.

Nevertheless, Odysseus's moral stamina is seen in his ability to forego false luxury, to refuse beds that are not his rightful resting place. For example, on return to Ithaca he first seeks shelter from Eumaeus, an old swineherd who has remained loyal throughout the king's absence. Eumaeus doesn't know who the hell this stranger is but, a model of hospitality, still offers him his own bed. Odysseus refuses and sleeps outside. Along the way, there are many accounts of Odysseus finding sleep and often these are a foretaste of his ultimate homecoming. Hospitality, kindness to travelers, and offering refuge are the timbers from which Homer built *The Odyssey*, a ship that is anchored to the wanderer's bed.

The point of *The Odyssey* is finding rest. It is curious that western storytelling, a treasure trove of restlessness, a vast anthology of itchy feet, begins with a tale whose substance is so different. Of course, there is a lot of seepage between the ideas of finding rest and death. But the hero of *The Odyssey* is a guy who has cheated death. He has come home not to die but to sleep. He has come home to bed.

9:45PM

[1997]

Once upon a time, before my wife, Jenny, and I got married and had our three children, I was a Catholic priest. On one occasion, I fell asleep during one of my own sermons, an accomplishment that is easier than it sounds.

I said Mass on Sunday evenings in a parish full of wonderful young families. I thought I was doing everyone a favor by keeping the sermon short, a discipline I achieved by sticking to topics I knew something about. Generally, my wisdom had petered out by the end of the third minute.

One day, after the service, a mother of three young boys took me to task for my brevity. The woman's problem with me was practical. She put good money on the collection and wanted better value. Mass was her only chance on the weekend to have a rest, and by late Sunday afternoon, she was totally exhausted and facing the weekly prospect of getting the lunches cut, the boys to school, and herself to work the following morning. The sermon was her only chance for a bit of a nap. Would I mind stretching it out a bit longer? She'd be grateful if I could. She needed the rest. I said I would do what I could.

The great Irish satirist and author of *Gulliver's Travels*, Jonathan Swift, was annoyed by people who fell asleep in church. He was an eminent churchman who made a career out of turning up his nose and being appalled. Swift divided the world into those who stayed home on Sunday to sleep there and those who went to church and slept there. He wrote a diatribe called *On Sleeping in Church*. It's hard to imagine the topic raising a sweat these days. Swift wrote,

> Of all misbehavior, none is comparable to those who come here to sleep. Opium is not so stupefying to many persons as an

> afternoon sermon. Perpetual custom hath so brought it about that the words of whatever preacher become only a sort of uniform sound at a distance, than which nothing is more effectual to lull the senses.

Swift had a point. I grew up hearing a story about a politician who used to come to our church. One time, he fell asleep during the service and began snoring loudly. His wife was embarrassed and dug him in the ribs to wake him up. Assuming he must have been sitting in the chamber, he immediately called out, "Hear! Hear!" The congregation was distracted because the priest had just begun to wonder aloud what words Jesus might have said to the deaf girl in that day's gospel reading—and "Hear! Hear!" was not a bad suggestion. The problem was that the priest was not open to suggestions; all the questions in our church were rhetorical.

Sermons are not the only cause of untimely sleep. Speeches, meetings, PowerPoint presentations, and children's ballet concerts can all have the same effect. The Internet has plenty of images of people who have fallen asleep at awkward moments, including a judge who dozed off during sensitive testimony in a trial for a violent crime. One poor anchorman fell asleep at his desk while on air. This may not mean it was a quiet news day. It could mean the opposite—that he was working too many hours, trying to keep up with the endless flow of wisdom that comes from the mouths of politicians and celebrities.

We inhabit a culture that keeps people on the brink of falling asleep and yet inhibits them from doing it properly. Swift wasn't really concerned just about church. He was concerned about people who were too exhausted by what was happening on the surface of the world to keep an eye on its foundations.

Heaven knows what Jonathan Swift would have thought when I nodded off during the sermon I was delivering myself. This may be divulging a trade secret, but once a sermon gets beyond a couple of minutes, it reaches a delicate point at which the preacher has no idea what he or she is going to say next. In the Jesuit tradition of which I was a part, this point normally came much closer to the start of the sermon than the end. The Jesuit custom was to keep the sermon ticking until something popped into your head, a practice known as "relying on the spirit." It was a risky way of going about things, especially when the most likely thing to pop into your head was either what you'd already said or what you'd soon wish you'd never said. Another strategy when stuck for an idea was to pause briefly and invite the congregation in a reassuring tone to reflect on what you had just been saying. This bought a bit of time to come up with something else. It was on one such occasion that, with my hands joined devoutly on the lectern, my head started to nod. My eyes closed. My breathing slowed and deepened. It was only when I bumped into the microphone that I woke myself up and noticed that the congregation was giggling. I remember thinking that I must have said something funny and wondered what it was.

(This wasn't the first time that I had amused a congregation when I thought I was being serious. Soon after I was ordained, I was asked to preside at a service on Good Friday, the day Christians ponder the death of Jesus on the cross, hardly the happiest moment in human history. I wanted to make the point that the message of Jesus was hard to reduce to a few choice slogans. Unfortunately, what I said was that Jesus was a hard man to nail down.)

At least now I had a bit more experience under my belt: it appeared that I could give a sermon in my sleep.

ZZzzzzz

The incident was part of a bigger picture. I wasn't just falling asleep during my own sermons; I was falling asleep anywhere and everywhere. I would go to bed early, get up as late as possible, and yet, by ten o'clock in the morning, there was nothing I wanted more than to go back to bed. After lunch, I'd crawl under my desk at work and grab some shut-eye. It was getting harder and harder to stay awake. At the same time, my snoring was getting worse and worse. I could make a monastery sound like a factory. I was like the Giant in Jack and the Beanstalk who rocked the entire castle with his snoring.

I underwent my first sleep study in January 1997 at the age of thirty-six. The cheerful technicians stuck a suite of electrodes to my scalp, chest, and legs and put a band around my chest to measure my breathing. They also put a microphone somewhere to record my snoring. (This, I presumed, was how they made sound effects for disaster movies.) The electrodes were all connected to an electroencephalogram (EEG) machine, which traces brainwaves, drawing a picture of what the brain is doing during sleep; this computer was kept outside in a command booth. Then, trussed up like a turkey at Thanksgiving, I was asked to get as good a night's sleep as possible. I knew at once that this was a ridiculous request. I wasn't going to sleep a wink. Nobody had advised me to bring my teddy bear.

When I got the results, I discovered it had taken me nine minutes to fall asleep, an interval known as sleep latency. Even in the comfort of your own bed, if you fall asleep in less than ten minutes it's an indication of sleep deprivation.

There were more results as well. The following morning the technicians came in with big smiles and said that they had big news to tell me.

"Oh," I asked warily. "What's that?"

"We can't tell you."

"How come?"

"You have to see the doctor in a couple of weeks."

"The suspense will kill me."

It wasn't suspense that was going to kill me. It was sleep, or at least what was happening in my sleep. When I turned up for my appointment with the doctor early on the day after a public holiday, his waiting room was packed, suggesting I was not the only person in the world with problems. Meeting the doctor, John, was one of those experiences—a bit like what I imagine it is to discover that your partner has been having an affair for years—when you realize that you have known very little about a major part of your own life. John produced an impressive little pile of printouts, technically known as a polysomnogram, that were generated during my night in the sleep lab. He started circling parts of them with a magnificent black Mont Blanc fountain pen that I began to covet.

"How do you think you slept in the lab?" he asked.

"All things considered, not too bad," I replied.

"Were you aware of waking in the night?"

"No, I reckon I slept right through."

"Undisturbed?"

"Totally undisturbed."

He was writing all this down with his Mont Blanc. He then put the cap on the pen with a small flourish, indicating it was time for him to stop listening and start speaking.

"Actually," he said, "you woke up 287 times."

I found this hard to believe. Perhaps he'd mixed up my results with those of a young mother somewhere. The Mont Blanc reappeared and circled the key statistic, to impress upon me that 287 was a very big number and not to be joked about. I noticed the pen had a broad nib, the type that requires skill to wield without making a mess, altogether a very nice writing instrument. I wished we could talk about that.

"You slept for a total of five hours and forty-nine minutes. This means you were waking up on average forty-nine times an hour or, in other words, almost every minute."

"What about the snoring?"

"This is related to the waking. I'll explain how that works in a minute. We were recording you at well over eighty decibels, which is the same as traffic noise or shouting. Normal conversation is sixty decibels; hearing damage starts at ninety decibels. You weren't far short of that. It was quite a racket, I believe."

"Wow."

"Yes, 'wow' indeed."

There were some important things I needed to understand. The 287 interruptions to my sleep were called *apneas*, a word of Greek origin that means the cessation of breathing, and I had a condition called obstructive sleep apnea, which, in those days, was not well known. That situation has changed dramatically in recent years, as the ailment has reached epidemic proportions in well-fed countries. Sleep apnea is a condition that more commonly afflicts men, although their partners also suffer. It does affect women themselves as well, although it is one of the few maladies for which men are more likely to go looking for help, usually with considerable urging from home. Dr. Christopher Worsnop, a sleep physician, explains that the classic interview with a couple goes like this:

Doctor: "Do you snore?"

Man: "She says that I do."

Doctor: "Does she snore?"

Man: "She says that she doesn't."

Being overweight is a significant risk factor. As a luggage handler once said, I had a bad case.

Sleep apnea is in large measure the result of a design fault in the upper airway. The human throat is a floppy tube, something that distinguishes us from all other species, which have rigid throats, a situation that is thought to have come about because of the human need to speak. While you are asleep, your tongue and soft palate, which is the fleshy part at the top rear of your mouth, relax and your throat collapses. Your uvula, which is the bit that hangs over your tongue like a stalactite and which you can see when you gargle, also flops in the way, as do your tonsils. As a result of so much slack behavior behind your teeth, the passage of air to your lungs may be blocked, especially if you've had a bit of alcohol or if your throat is narrow. Why might your throat be narrow? Perhaps you're a bit chubby: the body stores fat in visible places and also invisible ones such as the walls of the throat. On the other hand, it might just be a matter of luck. People with jutting jaws are more likely to have open throats and hence be less susceptible to snoring and sleep apnea. Dr. Worsnop points out that superheroes such as Superman and Batman are often drawn with strong jutting jaws, a feature that, since the time we lived in caves, has been seen as attractive to women. I personally think the reason women may be attracted to jutting jaws may have nothing to do with jutting biceps or jutting anything else; it simply makes it less likely that they will have to put up with snoring.

If your throat falls in on itself or becomes obstructed, the level of oxygen in your blood decreases and the amount of poisonous CO_2 rises. If something didn't happen at this stage, you'd suffocate. But luckily the increase of CO_2, decrease of O_2, and the work of various receptors in the throat, lungs, and chest all send a message to the brain that it needs to wake up and the brain obliges. The brain does a lot of things for you without even letting you know. It's good like that. The loud spluttering, strangling, gargling noise that passes as snoring is actually your attempt to push the palate and tonsils out of the way, open the throat, and clear the airway. The noise sounds desperate, and it is. You are struggling for life and you don't even know. Untreated sleep apnea is a killer; the main way it kills people is when they fall asleep driving. If you don't actually choke, it may put pressure on the cardiovascular system. Even if you avoid these pitfalls, you wake up exhausted, as someone who has been disturbed 287 times in a night has every right to be.

"Would this be happening every night?"

John picked up the pen and held it between his two forefingers like the rod of judgment.

"Yes. Every single night of your life. You're lucky we found out. It was five minutes to midnight for you."

I must have looked shocked.

"Don't worry," he said. "I love the old Cold War language."

Had I been talking with John twenty years earlier, his options would have been limited. I could have tried sleeping on my stomach, an old-fashioned idea that can make a difference because it allows the soft tissue in the upper airway to fall forward and make less of a nuisance of itself. The advice given to snoring blokes in a bygone age of putting a tennis ball in a sock and pinning the sock to the back of their pyjama top is not just an old wives' tale. Another idea is to put on a bra backward and put tennis balls in the cups, a form of evening wear that can be confusing to a partner in the middle of the night.

Upping the ante, he could have suggested a tracheotomy, an operation that puts a little hole (a tracheostomy) in your throat below the site of the blockage. This hole is then left open at night, like a window, to let some air in; but it is not a sightly addition to the physiognomy, as it makes a person look a bit like a bassoon. The air bypasses the collapsible throat but in so doing also bypasses the vocal cords, so you can only speak during the day by inserting a plug into the hole. A further and yet more drastic option may have been an uvulopalatopharyngoplasty, a word that required nothing less than the services of a Mont Blanc fountain pen to get itself onto a piece of scrap paper so that I could contemplate it with all its vowels.

"Don't worry," said John. "It is normally just called a UPPP."

A UPPP involves the removal of the tonsils as well as a serious trim for the soft palate, the uvula, and the pharyngeal arches, whatever they are. John wasn't recommending this form of major surgery. It tended to be very painful and was by no means guaranteed of success. Like a vasectomy, it isn't a procedure you can do yourself.

But luckily there was something that had become available of more recent times. It was called CPAP (continuous positive airway pressure) and was the brainchild of a professor in Sydney named Colin Sullivan, of whom John spoke with awe. Sullivan had come up with a clever solution to a problem that had baffled the boffins for ages. While others were dabbling in such elaborate ideas as injecting silicon into the soft palate to stiffen it up so that it maintained its condition during sleep, Sullivan realized that the upper airway is a bit like a door that keeps banging shut in the night. It just needs somebody willing to stand with a foot in the door. Sullivan theorized that what was required was a machine that would use simple air pressure to splint open the airway; the machine would fit into a mask, and the mask would sit over the nose of the patient. It was a simple but ingenius mechanical solution to a problem for which others had sought surgical, pharmacological, and even psychological solutions. Colin Sullivan's bright idea has saved tens of thousands of lives.

I returned to the sleep laboratory to experiment with CPAP, and the results were remarkable. John explained to me that sleep has distinct stages, each stage marked by a certain type of brain activity; the function or purpose of each stage has long been the subject of argument and conjecture. These stages rotate through the night in cycles of approximately ninety minutes; five cycles is a good night's sleep for most people. The fifth stage, which begins an hour or more into a night's sleep, is in a class of its own and is so unique and mysterious that it is often known as paradoxical sleep, meaning it is a time when the body looks asleep and the brain looks awake. It is called REM sleep, and it is compeltely different from the other four stages, which are known as non-REM, or NREM, sleep. Stages three and four have a particular importance and are known as slow wave sleep. Many sleep researchers don't divide the states of human wakefulness into sleeping and waking. They divide them into three separate categories: waking, NREM sleep, and REM sleep.

Like many people with sleep apnea, I was missing out on stages three and four sleep, the time in which growth hormone is released, a substance that uses small doses to achieve a long list of useful results. I was more than just tired. I was sick.

"You also have periodic leg movements." John Mont-Blanced this new problem onto page two of the polysomnogram. "This means your legs are moving all night long, kicking. You spend the night walking without going anywhere."

"Why?"

"Well, it's one of a number of sleep disorders in which people do rhythmic things during the night. Bruxism is another one. Teeth-grinding, in other words. It's most common in children and more common in adult women than adult men. We think these sorts of disorders may be some kind of release mechanism."

"I probably need the exercise."

"You're lucky you're a priest. At least nobody else is going to get kicked in the night."

He produced a cheap ballpoint that he kept for writing on a prescription pad on which carbon copies where required and wrote a prescription for a drug called pergolide (sold as Permax), usually given to people with Parkinson's disease to control their shaking. Permax was to become embroiled in controversy a few years later and was withdrawn from the market, angrily pursued by a group of people who blamed it for causing compulsive behavior such as gambling. It was also held responsible for heart-valve problems in people with Parkinson's. John didn't know that at the time. He just thought it might cause nausea, like motion sickness, a strange side effect for something that prevented motion.

"There's more to sleep than meets the eye," I said. "I never realized."

"Well, most of it happens in the dark, so it doesn't meet the eye, which is why it's been one of the last frontiers of medical research."

Over time, John elaborated. He told me that many people think sleep is a passive state.

"It's not like that at all. I think of it as the night shift coming in. The plant doesn't close down. There are all sorts of active processes going on that need to happen overnight."

"So why do we sleep anyway?"

John drew a deep breath. "Well," he said. "It's not like there's one explanation. It depends who you ask."

If you ask an anthropologist, human sleep evolved to keep our ancestors safely in their caves at night away from nocturnal predators. If you ask a neurophysiologist, sleep is when a lot of neurochemicals get replenished; in other words, it is when the brain eats. If you ask a physician, sleep has a metabolic function; it's when a lot of tissue repair takes place. If you ask a psychiatrist, it's all about memory consolidation and the reprocessing of information, and dreams have a role in this. If you ask a developmental physiologist, sleep may be a remnant of our fetal existence and could be a hangover of circuit-testing in the fetus when dreams and dreamlike activity are important for helping a brain discover what it can do and teaching it how to do its job. Most fetal sleep is REM sleep, the type in which the brain is really pumping. The percentage of REM sleep diminishes over a lifetime. If you ask an adolescent pediatrician, you will discover than in the months before puberty, the pituary gland is working double time during sleep to get the process started.

I interrupted him." What about you? What do you think it's for?"

"I think it performs all these functions."

"Really."

"And more besides."

The long and short of it is that no one fully understands why we sleep, but everyone agrees that sleep is both vital and universal. Fish, amphibians, and reptiles don't have REM sleep. Birds and mammals do. So it could be that REM sleep is a sign of having got further up the evolutionary ladder. Nevertheless, it appears that even insects have inactive and active periods, much like a sleep/wake cycle, although lab technicians have found it difficult to get those electrodes onto the brains of bees.

John wrote down the name of a chemist who sold the kind of CPAP machine I required, as well as the pressure at which the machine would need to be set. I didn't notice what pen he used. I was too busy looking at where I had to go next.

The best part of two decades later, in the early days of 2017, I caught up with John again. Both our lives had moved along in surprising ways, and his rooms now had a commanding view of St. Patrick's Cathedral in Melbourne. We shared a laugh about the ways in which the cathedral had been helping people to sleep for longer than he had. I asked him about the latest news from the world of sleep medicine.

"The biggest change is in the area of awareness," John said. "Everyone knows about sleep now."

I wasn't surprised to hear this. There was hardly a day when sleep wasn't in the news in some way, shape, or form. It was even beginning to compete with restaurant reviews as a topic for lifestyle discussion.

"People now see sleep as a component of wellness. They used to talk of the importance of diet and exercise. Now they talk about diet, exercise, and sleep."

Every year, John tries to attend the meeting of the American Academy of Sleep Medicine, which takes place in cities such as Boston and Denver and attracts more than six thousand registrants. These are just some of the hardcore professionals in the field.

John went on to describe how his practice had developed. "Seventy percent of our patients who have sleep apnea have one or more other sleep issues. We focus just as much on them. We have seen strange sleep behaviors such as parasomnias and even what we call sexsomnias feature in a number of legal cases."

These are situations in which people do things while asleep that they wouldn't dream of, so to speak, in normal life.

John continued, "Judges will now accept the fact that a patient has a parasomnia but may not necessarily accept that as a defense for a crime committed in the night."

Sleep still keeps John awake at night. After twenty-five years in the game, he is clearly still buzzing with news from the frontiers of research. He speaks, for example, about new understandings of the role of orexin (also called hypocretin), a neuropeptide, in regulating our sleep/wake function. This has had major implications for the treatment of narcolepsy and the creation of new drugs that are continually arriving on the market. There is an entire new class of drug called DORAs, meaning dual orexin receptor antagonist.

"There are multiple competing DORAs coming through. Those drugs are worth a lot of money," John says. "A lot of money."

This prompts me to mention the problems I had with Permax, the mischievous drug I had been taking for restless legs. John listens sympathetically. "Twelve of our patients who were using Permax lost a total of $12 million through gambling," he says candidly. When it comes to sleep, he treats drugs with extreme caution.

10:00PM

[1988]

Teachers tell stories about sleep, same as most people. Sometimes they are stories about all-nighters required to complete reports or mark papers. I would attribute some of my worst errors of judgment in the classroom to fatigue, although even the most alert people can still be foolish.

I recall speaking to a sixteen-year-old student whom I had been teaching philosophy for the whole semester. I was grumpy with him. He seemed disengaged and withdrawn, not making much contribution to a class that required students to chew on their ideas in public. I had taken a phone from him several times; each time it was like asking him to leave his mother. The problem was that I could not for the life of me remember his name. I was mortified, and he was rightly indignant. I used the wrong name. His breath shortened with impatience. I apologized. But still the name would not come. I didn't mention that I had been up at 1:00 AM trying to make sense of his assignment. He didn't mention that he was probably sacrificing sleep to his phone. Two tired people will never have much space for each other. Exhaustion can rant and rave and sometimes stamp its foot; it is seldom a good listener.

Philosophy is that rare thing, a necessary luxury. Plato thought that it belonged to the hours of the day when everyone was wide awake. In his time, that meant dawn. In our time, it's hard to imagine when that might be.

Other teachers tell very different stories about sleep. Martin Kelly has been a teacher since 1973. For the past twenty-five years, he has worked exclusively with young students who have been unable to find a home in what you might call traditional schools. These are often damaged children, kids who have coped with more in fifteen years than other people do in fifty. It can be difficult to gain their trust and create stability in their lives, but Martin is one of those self-effacing people whose quiet exterior belies a profound commitment. He is employed in one of the flexible learning centers under the auspices of Edmund Rice Education Australia. Martin's work is mostly in urban environments, on the sprawling edges of cities where supports and services are often slow to keep up with population growth. His world is different from mine. He tells me about a parent-teacher interview in which the child discussed was in some disciplinary trouble. This boy, call him Johnny, was using his phone too much. So maybe, in some ways, our worlds do have similarities.

"I spoke to the mother," he says.

"How did she respond?"

"She was annoyed with him. She said she was going to cut back his dope allowance."

He also speaks of twelve-year-old girls being given bottles of Jim Beam for their birthday.

"It's a cliché to say they grow up quick. But really, they don't get the chance to grow up properly."

On a number of occasions over the years, Martin has come across students who need to sleep with their shoes on. This is a powerful image. The kids need to be ready to escape from domestic violence in the middle of the night. Somebody—usually a man, often drunk or drug addled, more often than not the product of a similar childhood to the one being inflicted on the next generation—will turn up ready to do god knows what. The kids are expecting this; they don't even untie their laces. They flee as fast as they can. Sometimes their mother is with them; sometimes not.

Martin also speaks about finding kids in the middle of the night walking the streets. They will be in small groups, often families, an older one shepherding the others. They will have the blankets wrapped around their shoulders that they grabbed as they ran out the door, narrowly in front of the gale that is pursuing them.

"You can measure the chaos of their lives in terms of the chaos of their sleep," he says.

One of the basic signs of a coherent life is the knowledge of where you are going to sleep. To function as a human person, you also need an orderly idea of *when* you are going to sleep, but the *where* comes first.

In Martin's neck of the woods, many young people live transient lives. They often bunk somewhere in crowded homes. He has heard horrendous stories of young girls who share beds with older men simply in order to find some privacy, two square meters to call their own for a few hours.

Sleep can be a window on a dark world in every sense.

The children I teach generally know where they are going to sleep. But this doesn't always mean they are well rested. Sociologist Hugh Mackay, author of *The Good Life*, has conducted extensive research into the question of what makes people happy, and it's not what you think it might be. Instead of money or freedom or family, Mackay's discovery was that the key to happiness is self-discipline. It seems self-control is a far more secure foundation for happiness than self-belief. In his book *The Good Life*, Mackay writes:

> We may be sacrificing the very thing that will lead to the deepest sense of satisfaction: self-respect based on self-control. There is no shortcut to that and no amount of self-control will get us there.

As young people negotiate the woods of adolescence, the discovery of self-discipline is fascinating to observe and even to help foster. For some it comes more naturally than others, but a lot of young people are helped through a period of angst, uncertainty, and self-doubt by having a structure in their lives on which they can rely and in which they feel safe. Knowing when it's time to put out the light is the most simple and underrated part of this. A day, like a sensible meal, needs to have a clear beginning and a clear end. The times for sleeping and rising are among the boundaries that young people, as they mature, need to establish for themselves. Many rites of passage are justly celebrated. These include the first date, the first kiss, the first breakup, the first paycheck, and the first time behind the wheel of a car. But ideally, before all these, comes the important transition from parents telling you when to go to bed to making that decision for yourself and sticking to it. Finding order in a single day is as much part of the task of adolescence as finding order in the entire world.

My teaching career did not begin well. In 1988, I was twenty-six years old and had already spent eight years studying as a Jesuit. Having risen rapidly through the ranks, I was suddenly appointed deputy assistant dormitory master in a boarding school conducted by the order. There were also classes to teach, which was lucky for my students. I had all the answers. I had no formal teacher training, but that didn't matter. My self-importance would more than compensate. History was about to be put on the right path. Sadly, humility was the form of self-discipline I most lacked.

Before long, there were fireworks in the classroom to rival New Year's Eve. The racket in my rooms could be heard from one end of the building to the other. The whole situation was descending into chaos and farce, saved only by the good nature of the students whose sense of comedy was opportunistic but seldom cruel. One boy, George, had fled with his family from South Africa.

"You need the riot police," he suggested.

Chastened and confused, I sought the advice of colleagues. This was not hard, as in those days, the entire staff assembled at recess in the middle of the morning. They didn't come for the tea and even less for the cheap instant coffee, the curse of which was so legendary that an empty tin was used as a prop for the witches' cauldron in *Macbeth.* The daily get-together was essential for the exchange of information and news. We didn't have computers, and text messaging was even further in the future. So people had to talk to each other. They still do, of course, but mostly to complain about computers.

The first person I approached was Jack. Jack was hewn from old timber. He had been in the air force during World War II and had been a chronic insomniac ever since. This is a common manifestation of PTSD (post-traumatic stress disorder). We will hear more about it later. In some respects, Jack had never gotten out of uniform. He referred to the students as "the enemy." At the conclusion of recess, he'd join the end of the long line of staff waiting to tip their unpalatable coffee down the sink and announce to the stragglers, "Well, back to face the enemy." Jack was known to fight for both sides. When the deputy-head walked the corridor, he sometimes announced an air raid and got his class to take cover under their desks. Jack refused to put comments on reports. He couldn't see the point. The school insisted and eventually Jack had to give some ground, which was not his natural instinct. He simply wrote "G" for good, "VG" for very good, or "VVG" for very, very good. The underachievers got "POOR." No further embellishment was warranted. It was a far cry from the highly strung and meaningless gobbledygook that teachers are expected to produce these days. The language of modern reports is just one symptom of the culture of anxiety and exhaustion that has taken hold of education: teachers are required to generate more and more words to say less and less. Jack's pithy comments were created in the same impeccable handwriting as covered the board at the end of every math class he taught.

Each morning, Jack would announce to the staff room how many sleeps it was until his retirement—hardly the most positive approach to a job. At the time we met, the figure was still in the thousands. Years later, after Jack had retired, I ran into him in the coastal area to which he had retreated from the enemy. He said he missed school and told me how many days since his last class. Jack was a recovering alcoholic although hardly anonymous, making no secret of his membership in AA. He regularly invited some of the people from his AA group to come and talk to the boys. This made a significant impact. Many things change in education, but dealing with compulsive behavior, especially with regard to alcohol, is a constant, whether it is in the lives of parents, grandparents, uncles, aunts, siblings, or the students themselves. It was Jack who first told me that "the addict is emotionally absent," an insight I have worn to the bone with constant use. Sleeplessness tends to hang around with one of its dishonest buddies, addiction. You only have to look into a casino at 2:00 AM to realize that.

It is not just the presence of gambling and grog that makes trouble but also the absence of everything that these things replace. In my first year of teaching, I was advised by an alcoholic colleague, Tony, that the secret of his success was "a brandy before the first bell." The problem was that there was no secret and not much success, only a life half lived and a fine mind gone to waste on endless repetition of the same tired gags. He hid behind endless deception both of himself and others. It was Tony who first told me that teaching was a branch of show business. There is truth in this, but actors need sometimes to be able to take off their paint.

Jack was usually one of the last to leave the staff room after recess. So it was, lagging behind and trying to delay my own return to what I was beginning to think of as a battle, that I found myself asking his advice. He was never anything but direct.

"Always make sure you're the first in the room," he said.

"Really?"

"Yes. I have seen struggling teachers try to cut short the lesson by arriving late and leaving early."

This described me at that moment.

"Never show fear. Get in there early. It's your turf. Make sure you hold your ground. Never give an inch. It's your turf."

Even in the circumstances, this was too belligerent for me. I didn't want to win the battle. I didn't want to have a battle in the first place. "Thanks, Jack. I'll think about what you say," I told him.

"Never surrender," he said.

Surrender is, of course, the first and most powerful step of Alcoholics Anonymous. I wonder even now how Jack ever managed to take it. It is also the first and most powerful step toward sleep.

Next, I sought help from Cath, another math teacher. She was much younger than Jack, and while she took no nonsense from the students, she had a much warmer disposition.

"I'll give you one piece of advice," she said. "Never be late for a class if you can possibly help it."

My heart fell. I thought I was going to get the same machismo about holding your ground and defending your turf and showing no fear and all that stuff that won the war.

"I always try to be the first in the room," continued Cath. "And as each child comes through the door, I try to catch them eye to eye. Just for a fraction of a second. Because their eyes never lie. And as I catch each eye, I remind myself that every single person entering that room has seen different things since the last time I taught them. Every one of them has a story that they bring into the room with them. Perhaps they have forgotten their lunch, they may have just broken up with someone, they may have fought with a friend, they may have just got a part in a play they really wanted, they may have just been selected for a team or missed out on a team, their parents could be fighting, their grandfather could be dying. The list is endless."

She had me thinking.

"I teach math," Cath said. "I know the difference between infinite and irrational. The square root of two is irrational. The possibilities in the room are infinite."

I laughed.

"I remind myself at the start of every lesson that there is a wealth of experience coming into that room," Cath said. "Sometimes I don't know what it is. Sometimes I do. Sometimes kids need you to know their whole story. Sometimes school is a welcome break from the rest of life. But the classroom is not just about me. My job is to enlarge and enrich that experience. I am not there to please the kids or be confined to what they happen to be interested in. My job is to meet them where they are and try to take them someplace fresh."

I took Cath's advice, trying to catch the eye of every child as they arrive for the lesson. With each passing year, they seem a little more tired and frazzled. Part of the load that each person brings into the room is the burden placed on their shoulders by their culture. Foremost is the expectation that, from the age of ten or even younger, they will live frenetic lives. I have seen sixteen-year-olds nodding off at 9:00 AM in the morning.

"We all know where they are at 9:00 in the morning," a dear colleague once told me. "But you need to know who they are, not just where they are. And if you want to know who they are, you need some idea of what's happening in their world at 9:00 PM."

Fatigue narrows the moral vision of people and clouds their humanity. The truly exhausted can't see past the hands on the face of their watch. In *When Breath Becomes Air*, a beautiful medical memoir about life and death and the fine line between the two, Paul Kalanithi writes of trainees in surgical oncology who have so many sleepless nights that they even privately hope that patients who arrive on the operating table may turn out to be inoperable, that their cancer will be more widespread than anticipated and thus surgery will be unnecessary. Desperate young doctors would then be rescued from another nine-hour surgery that would otherwise be "stretching out" before them. Of course, when they wake up to what they are really thinking, these budding doctors experience "a gnawing, deepening shame," but fatigue eats the moral core. This is why it is of such concern to teachers.

The irony is that the very fatigue that is caused by having too many choices in the world ends up crippling the ability to make those choices on the basis of what Jonathan Swift would call "faith and reason." Swift was among the first to articulate the way that the habitually tired are at the mercy of unruly emotions and end up getting drugged on "an amusement of agreeable words." The 24-hour news cycle is both a cause and a result of exhaustion. So, too, are endless consumer choices and excessive time in the thrall of screens. Those who are saturated with information actually absorb very little, let alone think about what they hear. They make choices on the basis of cheap emotion for precisely the same reason that an exhausted professional eats junk food on the way home from work. They are too tired to sort proper ingredients into a proper meal. Teaching is about helping young people sort out the world in all its wonder, beauty, and complexity. An athletics coach needs to start with people who are fit. A teacher needs to start with people who are properly awake.

10:30PM

[1980]

When, my daughter Clare was ten, I was surprised how easily she memorized a good part of William Blake's "The Tyger." It's a poem that tends to stick to the memory; its rhythm does half the work for you. I was delighted to see her mind chewing on the textures of strangely shaped words. Sleep helps to create memory; learning things by heart helps the memory build muscle:

> Tyger! Tyger! Burning bright
> In the forests of the night,
> What immortal hand or eye
> Could frame thy fearful symmetry?

The author, William Blake, might have been surprised to find his work between the covers of a *Bedside Collection of Children's Verse.* His *Songs of Innocence* all have a dark tinge. Blake was uncomfortable with the materialistic explanations of the world that dominated the century in which he was born, the so-called Age of Enlightenment. He described it as a time when "the soul slept in beams of light," meaning that too much understanding of one kind can allow deeper kinds of understanding to slumber. In an 1802 letter, Blake wrote some of his best-known words: "May God us keep / From single vision and Newton's sleep."

Blake often took aim at Isaac Newton, a man who did an extraordinary job of approximating the whole of the physical world to a small number of mechanical laws. For Blake, the world was simply not a machine. He painted Newton as a naked man in a cave, transfixed by his measurements and drawings. All around him is darkness, but his single focus doesn't notice. He may as well be asleep.

Nevertheless, the sleep physician I met—Dr. John—had opened my eyes to the role of science in understanding sleep. There are indeed lots of things about sleep than can be measured and calibrated; my overnight sleep study had made this clear. Eventually, Dr. John sent me off to a chemist with a prescription for the CPAP machine, the device that would help me to breathe through the night by keeping my airway open. I hadn't realized that snoring, the stuff of comedy, had been for me a sign of a nightly struggle to stay alive. I literally fought against suffocation for hours on end. No wonder I was tired. Yet the solution was simple and scientific.

The chemist to whom I was despatched knew a great deal about the history of sleep. He immediately launched into an entire mythology about these new CPAP machines and the Australian who invented them, Dr. Colin Sullivan. He said that Sullivan was a recluse and a millionaire—all because he had a single bright idea, out of the blue.

"Everyone was trying to discover what to do about sleep apnea," the chemist told me. "One night, Sullivan was sitting in a Chinese restaurant in Marrickville. He was the last to leave, and they began to clean up the restaurant before he was finished. He noticed that the restaurant owner didn't sweep up the grains of rice off the floor. Instead, he had reverse-wired a vacuum cleaner that he used to blow the rice out the back door. That was when Sullivan had his Archimedes moment. What he came up with was basically a reverse-wired vacuum cleaner to push air down your throat all night to let you breathe." It was that device I was myself about to purchase.

I finally met Professor Colin Sullivan himself in September 2004. It turned out that almost nothing the chemist had told me was true. Sullivan had never been in a Chinese restaurant in Marrickville in his life, and he was far from a multimillionaire recluse. Intead, his office at Sydney University was still in the same unassuming room in the same building where he had devised the first CPAP machine almost twenty-five years before. He invited me to pay him a visit.

By the time we met, a million people around the world were using his invention, and it had become Australia's second-largest medical export after the cochlear implant for hearing. And yet Sullivan's office still didn't have air conditioning. When I arrived, he was struggling to get the window open.

"It gets difficult to breathe in here," he said.

It was breathing that brought Colin Sullivan into the area of sleep medicine, which, in the 1970s, was very much in its infancy. In many respects, it still is. Sullivan was drawn to medicine through physiology, the study of the more mechanical aspects of the human person, because it was the closest branch of medicine to engineering. His two elder brothers had both been engineers, and it seemed to be the family default position. His father had been an electrical fitter.

"My wife calls me a human engineer rather than a doctor because I tend to think mechanically. It's only recently that I have begun to think of myself as an inventor," Sullivan told me.

In the early seventies, Sullivan had embarked on his research under the supervision of David Read, after whom the sleep laboratory in which we were sitting is named. Read became interested in the tragic phenomenon of sudden infant death syndrome (SIDS) after friends had lost a baby in this way. The diagnosis was still in its early phase; SIDS had only been named as a distinct condition in 1969. Read began to explore the nature of breathing and sleep in infants. Over time, Sullivan moved on to asking separate questions about sleep in adults. He went to Canada and did research on what happens to the breathing of adults during sleep, using dogs fitted with masks to aid his research.

"We are pretty sure now that, in children, sleep drives the entire process of development," he says. "In adults, it has more a function of maintenance."

Ironically, after thirty years in the field, Sullivan is now more interested in childhood sleep and especially fetal sleep than ever before, developing techniques for studying it. He was trying to ascertain if there were clues in childhood, especially in childhood breathing and snoring, that might provide pointers for the later onset of serious conditions such as sleep apnea. He wondered if they could be headed off before they developed and did damage. He points out that it was once common to remove large tonsils from children.

"These kids often presented as sickly, and we used to rip out their tonsils at the drop of a hat," he explains. "Their snoring was interrupting their sleep and suppressing their growth hormone. After the procedure, they'd often have a growth spurt."

He went on to explain that babies can spend eighteen to twenty hours a day asleep. Most of that sleep is REM sleep, which is that part of sleep where the brain stimulates itself. But fetal sleep is even more predominately REM sleep. So what's going on?

"We know that by eighteen weeks, the fetus starts to perform the motions of breathing, even though it doesn't need to," Sullivan explains. "It gets all the oxygen and nutrients it needs through the placenta. In its last four or five months, the fetus looks like it is breathing a lot of the time, and this requires a good deal of energy, so there must be some point. What we know is that the fetus learns and practices three key activities while it is asleep: breathing, sucking, and swallowing. These are critical to survival at birth. They are learned during fetal sleep. So, like adult sleep, it isn't exactly down time. It is key learning time."

Colin stood up and reached to the top shelf of a bulging cupboard and pulled down a box in which he started rummaging, picking over pieces like it was a box of oddments in a garage sale. These were the first masks that had ever been used for CPAP machines; it was a little box of medical history. They were ugly and cumbersome. There were plaster-of-paris molds in the box as well, masks taken from the faces of early patients. They had all been made by hand and fitted to each patient individually. In the early days, they had to be stuck on every night with glue and then prised off in the morning. It was a hell of a business. But the people who were prepared to sleep with these hideous gadgets stuck to their faces were in dire straits. They were people for whom the most natural thing in the world—breathing—did not come naturally, at least not in bed.

To Lucy Costas, a young science graduate, Sullivan was an entertaining lecturer who, according to her, was "a bit of an iconoclast" who "didn't kowtow to the medical hierarchy." When Costas began working with Sullivan and his team in 1979, she had newly returned from overseas with her husband. Not sure of life's next direction, her eye happened to fall on a job advertisement in the paper that wasn't quite what she was looking for but that she thought might do for now. She soon found herself in a new world.

Costas remembers hearing about a particular patient from the moment she'd first arrived. He was in his fifties, had been a heavy smoker, and was overweight. He was what is sometimes known as a "blue bloater," a condition in which the body begins to tolerate low blood oxygen levels and high levels of CO_2. As a result, he had developed a blue physical appearance. And yet studies showed that there was actually nothing wrong with his diaphram; his rib cage was expanding and contracting as it was supposed to, so his breathing apparatus ought to have been doing a better job. But there was little airflow at his nose and mouth during sleep.

Medical science had believed that such a problem must be neurological; it had to be that the brain wasn't communicating properly. Sullivan and his team, however, clarified the issue. The problem was really the collapse of the upper airway; the mine shaft was blocked near the surface.

"Up until that point," says Costas, "respiratory medicine didn't involve the throat. It basically ended at the neck."

Initially, Sullivan had thought that patients with this condition were few and far between, and that he'd have to go looking for them. He wondered if he'd find five a year.

But they came looking for him.

Just as he had done in Canada, Sullivan used dogs fitted with masks to help him understand breathing and sleep. He used German shorthaired pointers because they were placid, easy to train, and had short hair, which made life easier when it came to keeping equipment clean.

One day in 1980, Sullivan was visited by a man who had been scheduled to undergo a tracheotomy. The man was in his early forties, had a young family, and had reached such a level of dysfunction—being scarcely able to stay awake at all—that he was willing to undergo the extreme procedure. Needless to say, he was not looking forward to it. He was visiting Sullivan because he was volunteering some "before" and "after" studies to measure the procedure's effectiveness, but during the process, he kept asking Sullivan if there were any alternative at all to having a tube permanently sticking out of his throat, as he would have with the tracheostomy.

Sullivan didn't have any ideas and then, all of a sudden, inspiration struck. Sullivan thought of the German pointers and an idea popped into his head.

Sullivan suggested that he could fit a mask to the patient and hook it up to a machine that looked like a reverse vacuum cleaner, similar to one which had been used on babies considered in danger of SIDS. And so, at 9:45 PM one night in 1980, having come in for yet another all-nighter, Sullivan trusted his wild instincts and fitted the mask.

By 4:00 AM the next morning, he could hardly believe what he had seen.

"It was incredible," Sullivan says now. "The first experiment just worked like a charm. We turned on the blower and this guy went straight into REM sleep and stayed there for two and a half hours. You never see that. A REM cycle might be forty-five minutes."

Lucy Costas says she remembers the excitement the following day. Sullivan was already telling people that they needed to have their findings published as soon as possible. Before long, five patients had similar experiences and were reporting dramatically improved daytime alertness.

So it was that on page 862 of the *Lancet* of April 18, 1981, written by the team of Colin Sullivan, Faiq Issa, Michael Berthon-Jones, and Lorraine Eves, there appeared an article modestly entitled "Reversal of Obstructive Sleep Apnea by Continuous Positive Airway Pressure Applied Through the Nares." (The nares are nostrils. The *Lancet* expects you to know that already.) The article pointed out that five patients who, without CPAP, had virtually no stage three or stage four sleep improved instantly once they tried it. It concluded that "the inherent simplicity and safety suggest that home use will be possible."

That last line should have been lit up with dollar signs. The publication in the open forum of the *Lancet* meant that it was now open season for anyone to turn the reverse vacuum cleaner into cash. The *Lancet* has an assiduous readership of both medical professionals and business people, two groups that are by no means mutually exclusive.

There were hurdles to get over yet. One was getting the medical fraternity to accept a simple solution, especially one emanating from Australia. Another was creating a machine that was user-friendly; the early ones were so big and loud that they had to be installed outside the bedroom and special pipes were required to bring in the air in the way an aquaduct brings water from a damn. Other issues involved finding a way for the air-flow to increase gradually to the desired level, so the patient could get to sleep before it reached full force. But far and away the biggest challenge was designing a mask that people could use comfortably.

Twenty-five years after Sullivan's discovery, ResMed—the company that began from Sullivan's work—occupies a vast site on the side of a freeway in Bella Vista, one of the western suburbs of Sydney. The ResMed plant looks more like a university than a factory. Near the entrance is the Healthy Sleep Center, outside which lie beds of lavender, an ancient ally of sleep; their presence lends a slight air of the esoteric to a place with a business edge. At the bottom of a gentle slope is the Innovations Building, which will eventually house three hundred engineers. Every room has access to the balcony that overlooks the Stream of Ideas, an artificial river that runs through the center of the site. Many of the battalion of engineers at ResMed are involved in creating the perfect sleep mask, a task that is more complicated than it sounds.

On the other side of the property is the factory itself. Within twenty-five years of Sullivan's first experiments, they were making two thousand CPAP machines and fifteen thousand masks every day—mostly for export and mostly to the United States, where ResMed competes fiercely with another company called Respironics, which does much the same thing for a similar share of the market. Only 3 percent of ResMed's trade is local.

The company's "core competency" is making masks from Silastic, a word that was manufactured by gluing together silicon and plastic. Twenty-six molding machines (soon to be thirty) churn out masks day and night. Every few months, a new development in mask technology comes into effect, usually a precise refinement making them quieter or less obtrusive or more flexible. Masks cost between $200 and $300 and need to be replaced regularly. I have a collection of my own discarded masks from the last decade: each one a small advance on the previous model, each one a bit more expensive. It's like a sleep mask graveyard.

After thirty years, Sullivan said that he still spent half his time thinking about the meaning and purpose of sleep.

"So why do we sleep?" I ask him.

"It's hard to know," Sullivan says. "We would have a better idea if we could observe what happened to people who didn't sleep. But it's hardly an ethical thing to deprive people of sleep."

"You must have some idea why we sleep," I say.

"It's like asking why we eat," Sullivan says. "The answer goes in so many directions all at the same time."

A pioneer in the area of sleep medicine, William Dement, is famous for his response to the same question. When asked "What is sleep?" Dement replied simply, "What is wakefulness?"

Waking, of course, can have traumas of its own. The fairy tales never tell you that: Briar Rose (if you like the brothers Grimm) or Sleeping Beauty (if you prefer either Charles Perrault or Walt Disney) wakes after one hundred years and everything's gorgeous. The prince finds his way through the thicket of thorns, plants the kiss we've been waiting for, and everyone in the castle rises and shines and gets on with life as if nothing has happened. Even the cook, who fell asleep as he was about to hit the scullery boy, gets to land his punch. The only misgiving is in Perrault's 18th-century version, in which the prince notices that Sleeping Beauty is wearing the fashion of his great-grandmother's era and that her collar is too high—but she is so beautiful, and no doubt well rested, that he loves her anyway.

Lucy Costas continued working for many years in the area of sleep apnea, patiently fitting mostly jowly men to the masks that put the wind back in their sails. She thinks back to one of the very first CPAP patients, someone who found waking from his long years of slumber a difficult experience. He awoke in a world that was not the world in which he had fallen asleep. Things had changed. It had been years since he'd been alert enough to notice.

"I often wonder what became of him," says Lucy. "He didn't continue with his treatment."

Apparently, the journey back from his long hibernation was too hard.

"As far as I recall," Lucy says, "he chose to go back to sleep."

10:45PM

[2004]

When I was a child, I often found myself unable to sleep. I used to lie awake and listen to the way the world spoke at night: cats prowling, birds nesting, dogs barking, windows rattling, doors creaking, wind blowing, rain falling, mosquitoes buzzing. It was only the mosquitoes that disturbed me, because I hated getting bitten and itching all night. The other sounds were amplified by stillness. I was never more a creature than at night, because that was when I listened to voices that weren't human. They didn't nag, instruct, advise, or warn. Nor did they express affection, praise, affirmation, or encouragement. In other words, they were different from the human voices in my life. They didn't make any claim on me, either positive or negative. I still love that aspect of night. It's when humans get to leave center stage and be part of the chorus.

There was not one reason why I was such a poor sleeper. But there are likely many factors that have played a part in my nocturnal habits. First, there were the mosquitoes. I shared a room with my brother, and the mosquitoes seemed to prefer me; they would torture me all night long. But there was also the fire: On five separate occasions before I turned nine, fire ravaged through the valley behind our house and we had to evacuate in a hurry. At least twice we were in our pajamas when we were rushed out the door. (Then Mum sent us back for our dressing gowns. We may have burnt to a cinder, but at least we'd have been decent. Mum subsequently became a great advocate of flame retardant nightwear, which was hardly a comfort.) Fire often recurred in my dreams; I would sometimes wake and smell for smoke.

In a lifetime of change—growing up, becoming a priest, leaving the Jesuit order after twenty-one years and getting married and starting a family—my issue with sleep was one of the main constants in my life.

The year 2004 found me living with my new wife, Jenny, in the tiny town of Gunning, Australia, population just five hundred. By that point, I'd been through the process of diagnosis of sleep apnea, bought a breathing machine, and been fitted for a mask. In the whole of this process, getting a mask that stays in place, doesn't allow air to escape, and doesn't cause blisters is the most arduous part.

I'd been using the mask for a while by the time Jenny and I got together, so I was now quite accustomed to sleeping with a large device on my face. I was, however, quite worried how Jenny would take to it. A breathing mask is hardly the most alluring item of intimate apparel. The alternative, I explained to Jenny, was the equivalent of sleeping beside an idling V12 engine. Without this ingenious invention, the product of exhaustive research in the style of Thomas Edison, we'd need not just separate rooms but probably separate houses.

Jenny was so good at overlooking the mask that before long we were expecting our first child, Benedict.

The local doctor in Gunning had a plaque hanging over one of her patient chairs with a message replicated from a time when electric light was still attached to the apron strings of its creator. It read this room is equipped with edison electric light, and it assured the reader that there was no need to put a match to the light and that electricity has no harmful effects. nor does it effect soundness of sleep.

I was studying these words with furrowed brow as our doctor told us, for the second time in two years, that Jenny should make an appointment to see the obstetrician. I studied them again when we returned to the doctor and told her that the obstetrician had squeezed some jelly on Jenny's belly, dug around in the jelly with something that was connected to a computer, and soon discovered that Jenny was expecting twins.

“That means two babies,” I had responded, dumbstruck.

The obstetrician replied that, yes, she did know what twins meant. She had read about it in med school.

The obstetrician looked at me strangely and then glanced at Jenny with pity. Twins were good news. A difficult husband was not so good. There were a lot of hearts beating in that little room, which was just as well because mine had stopped for a moment.

Jenny and I had plenty of time on the long drive back home to our town to talk things over. The obstetrician had gone though a long list of agonies that could lie ahead. Jenny was told she had a high possibility, even probability, of miscarriage, and if the babies held on to their slender thread, there were other possibilities that were also labeled as risks. By the time we had had something to eat and had passed the last street light of the city, it was already 9:30 and we were safely under cover of dark. It was then that we could both admit that we'd hate to lose either of these tiny creatures. No matter what was involved, we wanted them to come and live with us. I think the night sky of our country town had something to do with that, for big things seldom seem so big when you can see the stars. It doesn't always feel the same way in the city, a place where darkness has to be artificially created. It has no night sky to keep things in perspective. You can thank Edison for that.

We returned to our local doctor who, other than Edison Electric Light, didn't have a great deal of gadgetry in her surgery, not even a computer. When we were expecting Benny, the obstetrician would tool around with her ultrasound and tell us that the baby's length was precisely 14.5 centimeters. She would join her thumbs over Jenny's belly, reach around with her little fingers, then hold her hands against an old wooden ruler and tell us that the baby was approximately 14.5 centimeters. We were looking forward to surprising her with our news about the twins.

"Oh, I thought so," she said when we told her. She showed us the record where she had written, five weeks after conception, "Probably twins." She had made an educated guess based on Jenny's size.

"That means two babies," I added, still wondering if there was perhaps a little-known technical definition of the word that might involve fewer nappies and more sleep.

"I know," she said softly. "Yes, two babies." She must have gone to the same med school as the obstetrician. "You'll find that's quite a lot of babies," she said.

There were practical considerations to think about while we were expecting our twins. One was that Benedict, now aged sixteen months, was already bringing sweetness and light into our lives. Perhaps a bit too much light. He was progressing wonderfully, a prodigy in every area of accomplishment, a child to make the young Mozart look like a hack. He had already surpassed his old man in both wisdom and maturity. There was only one problem: The meaning of the phrase "a good night's sleep" eluded him completely. Almost as soon as the obstetrician found two heartbeats in Jenny's tummy, the prospect of two more sleepless babies entered our minds. The fact that we would soon have to find a bigger house seemed like a mere detail in comparison.

Soon afterward, Jenny's mother, Coralie, came to visit, and we took advantage of her vast experience as a babysitter to go to the movies. We looked in the paper and found there was a 10:15 PM showing at the theater down the road. We live on a long road. Times were, in another life, when going to the movies was no big deal for either of us. We both went a lot back then. It was something you could do on your own without fear of being pitied by some couples or envied by others; in those days, a 10:15 PM movie was where we were most likely to run into friends with whom we'd go to supper afterward and drink strong coffee. Sleep was a resource we could just squander. Now the nearest movie house was thirty miles away and a 10:15 PM session felt decadent.

There were numerous advantages to living in a small town. Gunning now had a population of 504, although most afternoons, you'd wonder where they all were. It wasn't long before we were telling our neighbors and friends that, once the twins arrived, our little family would constitute 1 percent of the population. There would have to be 80,000 of us to have the same demographic impact in New York, and it's hard to find an apartment that size in Manhattan. A town that never sleeps is, of course, one that never wakes either. Our village didn't have this problem. It dozed on and off in its patched pajamas, stirring every now and then to remark that things weren't what they used to be. I was looking forward to days ahead when our 1 percent could swing a close vote on some crucial municipal issue, such as the size of garbage bins or pool opening hours.

The other advantage of living beyond earshot of a city is that Jenny and I got a lot of time to talk in the car. We left Benny still wide-awake with his grandma to give ourselves time to get to the late screening.

"What's the name of the movie?" Jenny asked on the way.

"Sorry. I didn't check."

"I just hope it's not too noisy. I need the sleep."

Our talk soon turned to Benny and how we were going to get him to go to bed more willingly. We had tried everything the books and some well-meaning strangers told us; none of it worked. The latest advice we'd been given urged us to establish a clear and calm bedtime routine, something like a ritual, so that Benny would learn to recognize signals that the day was ready to close for business and he might kindly now make his way to the exit. If the ritual included quiet things he enjoyed, such as reading books and saying good night to his numerous teddies, then, we were led to believe, we would find that Benny welcomed the end of the day. Benny might not welcome it, but we certainly would.

The film we saw that night was about the writer J. M. Barrie, creator of *Peter Pan*, the boy who never grew up.

At one point in the film, the actor playing Barrie says, "It's a terrible thing to put a boy to bed, because when he wakes up, he will be one day older."

Jenny and I both started sobbing.

When we got home, we found Benny still awake, but grandma sound asleep in front of the TV. She had had five children of her own. She knew what to do.

We nursed Benedict for a while, and we talked about the kind of world our three children would grow up in. We wondered what the future would hold for our little ones, both Benedict and the two on their way. We knew they would need us to hold them. But they also needed to be cradled within a culture, a civilization, a pillow of language and ideas on which they could lay their heads, a blanket of beliefs that would shelter them from the cold.

11:00PM

[2005]

helter our forthcoming twins, we
in all kinds of places within a
lly found one around the corner
heat of summer, with an army
hicles, we managed to move
ouse to the next in a single

restless, so I took a stroll
live trees near the fence;
breeze. Since Homer's
ility. Planting them is
s soon as you arrive
rwise, but you know

rse.
of humans who
d because they
an they do in
der to rest, a
full stride.

are able
cogni-
other
than
act,
ch

is that if a

res with few
from the top
relationship
plains, able to
nt. Other crea-

me in the open;
notice. Humans,
hy they are hap-

three hours a day.
es into place so that
waking themselves
ment of tendons and
s front legs when the
legs, the horse has to
ccasionally, it is unable
nce asleep, a horse may
erent level of sleep not
will be hearing more in
has such a fine system of
remain standing for years
ey are beautifully still.
r a day spent moving. I left
eturned to the house where
ns in the middle of a thicket
or wrapping.

them in the act. The only rule of thumb for the uninitiated
giraffe looks like it is asleep, then it almost certainly isn't.

Giraffes behave like this for good reason. They are creatu
enemies, mainly because no one else wants to get their food
shelf. So giraffes have been able to develop a cooperativ
with other species. They are the watchtowers of the open
spy trouble and warn others at any hour of the day or nig
tures can sleep because giraffes don't.

Horses are similar to the extent that they are also at h
they sleep like recoiled springs, ready to move on short
on the other hand, learned to sleep in caves, which is w
piest if nothing disturbs them before daylight.

Horses don't sleep long hours either, maybe two or
But they have developed a method of locking their kne
they can sleep on their feet without losing balance o
by falling over. This "stay apparatus," a canny arrange
ligaments, turns itself on automatically in the horse
horse starts to relax. To activate the lock in its rear
wiggle its hips until certain bones hook together. O
to get itself unhooked and the horse can't move. O
then lie down and, when lying, will enter a diff
unlike the human REM sleep, about which we
the small hours. But in general terms, the horse
resting on its feet that horses have been known to
on end without growing weary. When asleep, th

The sight of a sleeping horse is a tonic afte
the horse and the slips of our new olives and
I finally found Jenny, slumped on some cushio
of packing boxes and newspaper we'd used

She smiled at me.

"Feel this," she said.

I put my hand on her tummy.

The babies were both awake, and I could feel them kicking for the first time.

11:20 PM

[28 BC]

The Odyssey, originally sung in Greek, has a Latin counterpart. Virgil's *Aeneid*, still unfinished when the author died in 19 BC, is also a classic quest story. Its hero, Aeneas, survives the fall of Troy and, like Odysseus, sets out on an epic adventure under the guidance of a divine being. Aeneas's life coach is the goddess Venus, who also happens to be his mother.

The Odyssey and *The Aeneid* are chalk and cheese. If *The Odyssey* is a book about getting home to bed, *The Aeneid* is a book about getting out of it. In that distinction lies something essential about the difference between Greek and Roman culture. Greek culture finds its most natural expression in philosophy; Roman culture finds it in law.

We inhabit a culture that wants to tip us out of bed, to find our purpose in the big wide world, to make a mark. Greek philosophy has a stronger sense that purpose or meaning is part of the human hardware and comes free with every new person. It has a predilection for finding interior motivations, for seeking out a kind of harmony or balance or equilibrium that is precisely what many people mean when they use the word *home*, a far more resonant concept than a little piece of land buried under a big mortgage. In *The Odyssey*, ultimate purpose is found behind closed doors. In *The Aeneid*, it is the opposite.

The CliffsNotes version of *The Aeneid* runs something like this: Aeneas, a Trojan, survives the destruction of his city and sets sail, away from his home, to fulfill his sacred destiny, the establishment of an entirely new place called Rome. On the way, he and his crew encounter a storm, get shipwrecked, and end up on the shores of North Africa. Here Aeneas meets Dido, the founder and queen of Carthage, and despite her resolution that her first marriage to Sychaeus was more than enough for one lifetime, she falls in love with Aeneas. (Sychaeus had been killed by Dido's jealous brother. Nice family.)

Take a look at the way the story is told. We don't get the gory details of the last days of Troy, complete with Trojan horse, until Dido invites Aeneas to talk about it. One night, he agrees to open up about what he has been through, despite the fact that the hour is late and the stars themselves are starting to blink as though they can hardly keep their eyes open. Aeneas is part of a long line of storytellers whose purpose has been to cheat sleep of her due.

Often, English translations give the impression that Aeneas delivers his great story from a dais or stage. Really, he addresses his audience from an elevated bed. Once again, there is a long tradition of this. The bed, especially the deathbed, has often been used as a pulpit, a place from which to deliver ultimate truths, messages of sacred significance. The bed is the place for famous last words. On the day he died, Thomas Edison, for example, sat up and said, "It is very beautiful over there." I'd like to think he was not so much dying as finding another project to work on in the next realm.

History does attest to the fact that some of the world's great speechmakers reserved their most paltry efforts for their final scene. Winston Churchill's last words were "I'm bored with it all." His ally, Franklin Roosevelt, the man whose fireside radio talks sent Americans to bed under a blanket of thousands of words, died of a cerebral haemorrhage, slipping out on the tail of just five words: "I have a terrific headache." Karl Marx died after he announced that "last words are for fools who haven't said enough," a criticism that certainly didn't apply to him. In 1886, an unknown spinster called Emily Dickinson left the world before anyone had really noticed she was on it. As she passed, she said, "I must go in. The fog is rising." J. M. Barrie, the creator of Peter Pan, said, "I can't sleep" and then died, which fixed the problem.

Aeneas also used his bed to great effect. It was more than a pulpit. In his case, it was a springboard, the place to launch his mission. The word that Virgil uses to describe the place from which Aeneas narrates the fall of Troy is *torus*, a slightly unusual word, the same word he likes to use for the marriage couch, a word with formal sexual connotations, as well as the funeral bier. Ironically, the usual Latin word for *bed*—the word Virgil doesn't use here—survives in English as *lectern*, a place for reading rather than sleeping (although it can be both if your book isn't much good). Aeneas's storytelling, which is pretty gruesome at times, is an act of unwitting seduction and Dido is impressed. She is wounded, set on fire, poisoned, ignited, driven to madness, afflicted, et cetera. Virgil is in full flight when describing Dido's emotional state. For a minute, she sounds a bit like Troy, which Aeneas also left smoldering.

Aeneas and Dido get it together, at least for a while.

But Aeneas's mother is not happy.

It's hard to argue when your mom is divine. Venus has a fight with Juno, Jupiter's wife, who wants the lovebirds to make their nest in Carthage. Juno is present throughout *The Aeneid* as the goddess who wants to mess up the plans that fate has drawn up. Venus is always going to win. She is the mother-in-law from Hades.

Back on Earth, Aeneas can't sleep. At the behest of his mother Venus, Anchises—the shade of his father—appears to him to tell him what to do. (Being married to a goddess, it seems, is no easier than having one for a mother.) Venus also gets Mercury—the god who, as one translator puts it, "gives sleep and takes sleep away"—to intervene. Aeneas must wake up, get on his way, leave the emotional Dido in her proper place, and establish the city that will, in turn, establish the rule of law over the earth. There is more to this than callow duty. We are watching a tussle between desire and law, and Virgil plays it masterfully. When Aeneas tells Dido that it is time for him to go, she turns on a scene and asks if he has time, at the very least, to leave her with a little Aeneas as a memento. He refuses, and she collapses on her bed. Aeneas receives one of the coldest endorsements in literature, one whose overtones of family, destiny, virtue, and loneliness are hard to convey in English: "At pius Aeneas." *But faithful Aeneas.* It's hard to translate the flint-like quality of these words. They turn passion to stone. Aeneas was always going to fulfill his destiny, but he left it until the eleventh hour.

What does Dido do? At night, after Aeneas has left, while Virgil sings a hymn to "gentle sleep" (the time when, in the rather florid translation of the 18th-century poet John Dryden, "Peace, with downy wings, was brooding on the ground"), Dido alone is awake. Even her insomnia is pictured as defiance of the laws of nature. She builds a fire and puts on it the clothes and weapons that you-know-who has left in her bedroom. Finally she throws the fire on the very bed she has recently shared with Aeneas. Virgil makes a point on a couple of occasions here of using the same uncommon word—*torus*—for Dido's bed that he chose for the platform from which Aeneas delivers his seductive story.

Dido is not just burning her bed but also Aeneas's place of rest. She calls down a curse on him and his people, an act of bitterness that legend supposed was the ultimate source of the war that was later to string across the Mediterranean between Rome and Carthage. Then she runs herself through with a sword and throws herself on the fire.

Beds are significant props in both *The Odyssey* and *The Aeneid.* Odysseus's bed, carved from the trunk of that olive tree, is immovable, a place of stability, identity, and rest. In that epic, identity and rest are kindred ideas. But in *The Aeneid*, the bed gets burned, and Dido incinerates herself along with it. There is no rest for the good.

Among Aeneas's descendents are the most famous twins in history:

Romulus and Remus, the legendary founders of Rome. Their mother, Rhea Silvia, was a vestal virgin but, don't worry—their father was a god, so she did not compromise her calling to chastity. But virgins who got pregnant could lose their lives. So the twins were put in a basket and sent down the Tiber, in a vivid case of being driven out of home. They were discovered by a she-wolf who suckled them and looked after them before a farmer intervened and restored them to human community. The story goes that when the she-wolf found them, their basket was safely caught in the roots of a fig tree. I like to imagine they were sound asleep.

The legend of Romulus and Remus is hard enough to believe in its entirety. But if you add the possibility that twins could be happily asleep at the same time and in the same place, it becomes the stuff of sheer fantasy. In my experience, it's easier to believe the part about the breast-feeding wolf.

11:45PM

[2005]

Once Jacob and Clare arrived in our lives, we soon found that lots of people had lots of opinions about the best way to look after twins. This is understandable because there are more and more twins around. The incidence of twins in western countries has almost doubled since 1980 and now accounts for about one in eighty births. This can be attributed to more than IVF; it's also related to the fact that mothers are older these days, and even without medical intervention, an older woman is more likely to release two eggs in one cycle. I gather this is what happened to us. (Not that Jenny was old. I never said that. Let the record show that she wasn't old then and she never will be.)

In the months leading up to Jacob and Clare's birth, I read everything I could find. I was convinced that having twins wasn't going to be so difficult after all. We traveled to the nearest big city to buy a pram and chose a serious piece of equipment that had been designed to the millimeter to fit through most doors, so long as you remembered to keep your knuckles out of the way. The woman who served us had been selling prams from the same store for over twenty years and had customers coming back whose mothers had dealt with her in the past. The expectant grandmothers stood to one side and listened while their daughters asked all the questions they had asked a generation before, thinking they were the first person to ever ask them. The woman said that when she started, a twins' pram required a special order. Now she had four or five right there in the shop for us to choose from, a tribute to the increasing prevalence of twins as well as to the fact that people who have families later, even if they don't have twins, may well have two children in a period of twelve months and they will need to be wheeled around together.

"Why this one?" I asked about the model she had selected for us. There were others with cup holders and even one with a facility for an MP3 player so that the babies could listen to Baby Einstein or Baby Mozart or something else that might give them an edge in life. I wondered why the woman wasn't recommending one of these funkier versions.

"I get lots of feedback," she said, "and the babies sleep best in this one."

We had the credit card out of its holster before she needed to get to any other reasons.

We spoke to friends who had twins. They said the first six months would be unforgettable.

"Why?" we asked.

"Because you'll be so exhausted, you won't be able to remember a thing about them," they told us.

But it wasn't just the first six months. It was the first twelve months, a strange time that we longed to get through and at the same time wished would last forever. We wanted these two little bundles to grow up fast so that they could help themselves and let us get some sleep; at the same time, we hated the thought that they would ever change and stop needing us. It was a time of vulnerability for all concerned.

Whenever we were out, people who had twins approached us from all directions, wanting to remember when their own were little. We met parents whose twins were three, seven, fourteen, sixteen, twenty-four—you name it.

"My twins are fifty-five," said one woman. "I've been looking at yours and thinking back. It's a beautiful memory. The time you have now is so precious. Don't worry. Life will get easier. But it won't get any better."

On another occasion, Jenny was struggling to get three kids out of the car and into two prams. A woman came along and put money in the parking meter for her.

"I had twins, too," said the woman, her kindness like the touch of a feather. She vanished before Jenny could burst into tears.

In addition to the shows of support we experienced when out and about, we recieived more than our fair share of unsolicited advice. In my early expeditions with Jacob and Clare, for example, I was told that they should be wearing hats. I would normally have agreed heartily with this suggestion except for the fact we were inside when it was made. A woman loomed up at me out of the dog food aisle. She had read something about the effect of fluorescent light on babies and was concerned.

"Cotton hats are fine," she said. "Just plain cotton will do."

Like the ghost of Hamlet's father, she delivered her message and vanished.

This was so odd that I had to ring Jenny and let her know. While I was on the phone, another person interrupted to tell me that it was dangerous to talk on a mobile while pushing a shopping cart and that kids required my full attention, and if I wasn't prepared to give it, then why did I have them in the first place.

"Technically, it was my wife who had them," I replied, passing the buck as usual.

Another time, a woman noticed I was wiping their noses with an upward motion instead of a downward motion. She was concerned that I was pushing germs back up their noses and felt it incumbent upon her to let me know. Another time, a man came up behind me as I was putting the kids in the car because he thought I wasn't fitting their restraints correctly. On a different occasion, I was sitting in a coffee shop and gave Clare—then almost nine months old–some of the froth off the top of my cappuccino, the part with a bit of chocolate sprinkled on it, as a small treat. But I chose the wrong moment. The woman at the next table went into action. Did I have any idea what caffeine could do to a child that age? Was I so stupid that I thought the froth on top was free of caffeine? Did I have any appreciation of what caffeine could do to the cognitive development of a child? Did I realize that sleep deprivation was an officially recognized form of torture? I should have said that I was the one being deprived of sleep, but I was simply too tired to come up with the retort at the time and only thought of it at a quarter to twelve at night when it was too late.

The one piece of advice we heard over and over was to never sleep in the same bed as the kids. And I can promise that we did what they said. We may have been in the same bed but we never slept a wink.

11:59PM

[350BC]

Aristotle is one of those people who had an opinion on absolutely everything, including sleep. He thought that the hour leading up to midnight was a time of special calm, especially the minute before. This was because earthquakes were most likely to strike at midnight; failing that they would strike at midday. His meticulous observations of the natural world often led him to advance with supreme confidence toward some strange conclusions. He likened the earth to a living body and earthquakes to the exhalation of wind. So a seismic tremor was a bit like someone farting in the middle of the night.

Aristotle is seldom on anyone's list of favorite writers. His books have a hard edge to them: he was a collector of data, and in so far as he wanted to make a big picture of life, the universe, and everything, it was going to be made out of this raw material. He was an intellectual bower bird, and his philosophy was built like a nest, twig by twig, scrap by scrap. Aristotle covered a vast field, but in any area, from logic to biology, from ethics to cosmology, his work is meticulous but generally devoid of human warmth or storytelling. He engages the issue not the reader. It is curious that Aristotle chose the line of work he did because he didn't suffer fools gladly. Philosophy is the wisdom of suffering fools.

Perhaps Aristotle's turn of mind owes something to the fact that his father was a doctor in an age, the 4th century before Christ, when the most helpful thing doctors could do was tell you what they saw. So the boy grew up with an appreciation of the power of observation. For Aristotle, the whole natural world was a book of symptoms. If you watched them closely enough, eventually you'd be able to describe something like a general condition.

Just as significant in turning Aristotle into Aristotle was the fact that he went to a good school. The Academy was an institution that is said to have started in an olive grove, a place of rest, once sacred to the goddess Athena. Athena was the girl who steered Odysseus back to his bed.

The Academy was founded by Plato, and by the time Aristotle arrived there from Macedonia at the age of seventeen (around 367 BC), it had been going for about twenty years. The Academy has since given its name to a dizzying number of places that are nothing like it, least of all the one that hands out awards to badly dressed actors with lots of people to thank. The original Academy didn't have a syllabus, a prospectus, an enrollment policy, or a policy on the availability of nuts and chips at the tuckshop. It didn't have resources, facilities, mission statements, values documents, or projected outcomes. Nor, despite popular belief, did it make students think. On the contrary, it allowed them to think. There's a difference. The Academy didn't want to change the world, only to understand it. As a result, it changed the world.

Aristotle studied under Plato until the latter died in the middle of the century. Plato's output diminished after his death, which was rare in those days. It was not the case, for example, for Plato's own teacher, Socrates, who, thanks to Plato and his dialogues, did some of his best work posthumously. Once Plato was dead, Aristotle then spent the rest of his life trying not to be Plato. He returned to Macedonia but later came back to Athens and started his own school, called the Lyceum. The Lyceum also began in a garden and students did their philosophy by walking around, thus coming to be known as the Peripatetics. They preferred a restless philosophy, one of inquiry rather than contemplation.

Plato and Aristotle have come to be seen as opposites, bookends on the shelf of Western thinking. And they had different ideas about sleep.

It is not surprising that Aristotle should write a treatise on sleep. Sleep has always been widely practiced in philosophical circles. When I was at university, tutorials in philosophy were always held after lunch, when the circadian rhythms of the body tend naturally to dip toward sleep. The cultural practice of the siesta originates less in a need to escape hot weather and more in a willingness to honor what the body wants to do anyway. I discovered a long-held belief among undergraduates that the purpose of philosophy was to create rest, not disturb it. Students soon learned that the most comfortable way to sleep in an upright chair is with your hand placed on your brow like a visor, keeping the light out of your eyes and at the same time making it hard for others to know if they are open or closed. Almost as effective is the practice of resting your elbow on your knee and then putting your chin on your hand, furrowing your brow in a parody of concentration. This is the posture of Rodin's famous thinker, a statue that, when you look twice, appears to be an image of a man engaged in the futile struggle of humanity in the face of existential confusion—that is, the struggle to stay awake after lunch. With a bit of practice, you can be fast asleep but still look like you are deep in thought, which may indeed be the case. The mind can be more active asleep than awake. And more creative, too.

When I was studying Philosophy 101, one of my fellow students spent the entire course asleep. He always arrived before the rest of us and remained behind when we left, so some of us wondered if he lived in the room that had been occupied by our teacher, a near contemporary of Aristotle, since ancient times. The student seemed so comfortable that, at the start of the year, some of us had thought he might be the tutor.

This impression was underlined by the fact that Ann, the real tutor, a woman of some standing in the world of philosophy, also spent the whole year fighting sleep. She was a pacifist. So she didn't fight very hard. In her case, however, sleep was a pedagogical technique. She had found that the best way to keep a class on its toes was to doze off yourself.

While the professor slumbered fitfully, the rest of us talked about anything and everything, whether or not it was relevant to the lecture we had heard before lunch. That's the beauty of philosophy. There's nothing that's not on the course. Philosophy is rather like French cooking. It requires a lot of raw material. In my experience, French cooking involves a whole morning of shopping. Tired and cranky, you get home with a carload of food and then spend a long afternoon in the kitchen. After a ten-hour day and a trolley load of groceries, all you end up with is half a cup of gravy.

Our professor believed that first-year tutorials were mainly the shopping and chopping part of the exercise. She slept while all the great issues of life flew around her ears in a frenzy of adolescent self-importance. She'd heard it all before, many times, especially the learned opinions about sex delivered by students new to the game. About the third week of the term, she did intervene to ask if someone could lend her some coins for the parking meter. Then, at the end of the course, she announced that those of who came back next year would be ready to start real philosophy. The recidivists discovered that that was when the arduous work of distillation began. Philosophy goes through a lot of junk in the search for an essence. But you need the junk to start with.

Only one person in our tutorial group failed to speak all year. He just slept. He didn't even stir when one of the young women in the room, an advocate for "reading culture," "reading society," "reading movies," "reading fashion"—reading, it would seem, everything except books—posed the question of whether or not he really existed. To be heckled by a girl on the grounds of nonexistence would have excited a reaction from most males. But not this guy. He was an ideological sleeper, and nothing was going to budge him from his position.

Some of us were a little surprised when the gentleman in question topped the year. Perhaps he had the only mind that had remained unsullied by use; perhaps he had discovered sooner than most that the best philosophy is generous with thought but mean with words. Meanwhile, the young woman who wondered if he existed or not had gone out of her way to get proof one way or the other. She was pregnant by him by the time the results were published. She told her friends that the only way to get him to wake was to sleep with him. She went on to write a thesis in praise of Platonic relationships.

Aristotle was a great one for counting stars, Plato for looking beyond them. If Plato said one thing about sleep, Aristotle was bound to say the opposite. It's not that sleep was the biggest problem either of them encountered. But sleep is one of those issues in the history of philosophy, perhaps because it is so commonplace, that helps brings basic attitudes into focus. People reveal a lot of themselves when they start talking about sleep and even more when they tell you why they can't sleep.

Take Descartes and Hume for example, often seen as sitting in opposite corners of the boxing ring of the history of philosophy. This may not be the kindest analogy: philosophers fight without gloves. René Descartes was a man of profound doubts, which he expressed as certainties. Meanwhile, the Scot David Hume (who wrote in the 1700s) was a man of certainties that he expressed as doubts.

Descartes didn't like getting out of bed in the morning, seldom did so before midday, and would, in the end, believe that the insistence of his patroness, the Queen of Sweden, to see him one morning at 5:00 AM gave him the cold that cost him his life. Whenever Descartes did drag himself out of bed, he wondered if he was still asleep. His famous dictum—"I think, therefore I am"—arose in answer to the conundrum "How do I know that I am awake right now and not dreaming all this?" His response was that famous use of the vertical pronoun: if there is an *I* in the driver's seat, then you are awake and conscious. Descartes was using the notion of thought in a broad way. His position is not refuted by pointing to people who don't think but still seem to exist.

Descartes was known as a rationalist because he thought philosophy began in the mind. Hume was an empiricist because he thought it began with experience, with things that can be empirically measured. His dictum was more "I am, therefore I think." Hume was one of the more affable people to overturn the Western mind: Immanuel Kant said that reading Hume "interrupted my dogmatic slumber." Hume's astringent thought was at odds with his own pleasant character. *Le bon David* wanted people to be glad to know him. This was despite the fact that his radical skepticism led him to conclude that it wasn't really possible to know anyone. Or anything. Hume believed that just because the sun had risen millions of times before, there was no logical necessity for it to rise tomorrow. A thousand examples of a connection between two things could never be extrapolated to entail a causal connection between them. So, if Descartes got out of bed wondering if he was still in it, Hume got into bed each night with no philosophical conviction that he would ever get out of it. One day, he didn't. He died in full possession of his doubts.

For Plato, doing philosophy was a bit like waking up well rested. Dawn is a key image in his work. One of his dialogues, *Crito*, deals with the impending execution of Plato's mentor, Socrates. The relationship between Socrates and Plato was, incidentally, anything but platonic. This dialogue begins with the wealthy Crito visiting Socrates in prison at dawn. Crito says that he has been watching with envy while Socrates slept peacefully. Unlike Crito, who has been unable to sleep, Socrates is represented as a figure of order and moral calm. His integrity is reflected in unblemished sleep; his clear-sightedness is evident at dawn. Similarly, in his final dialogue, *The*

Laws, Plato portrays an ideal city called Magnesia, a name that implies greatness. Plato enjoyed toying with the concept of an ideal state. He was the first writer to describe the lost continent of Atlantis. In *Laws*, the supreme authority in Magnesia is a "council of the night," a gathering of elders and philosophers that meets every day between first light and sunrise, the time when members would be most wide awake and least preoccupied by mundane affairs. Fresh from sleep, this group would discuss ideas. Plato's idealism is evident in his belief that, if you jailed prisoners within earshot of this meeting, their behavior would be corrected simply by hearing about The Right Thing To Do.

In Plato's ideal world, philosophy is the work of dawn. He distrusted night. In his best known work, *The Republic*, Plato discusses sleep in the same breath as describing the "tyrannical character" that forms part of an "imperfect society." He says that sleep incapacitates "the reasonable and humane part of us" and allows the "bestial" part to do its worst. In other words, he is suspicious of dreams, which lack both "sense and shame." The answer is this: as you get near to sleep, discipline your mind and feed your reason with "intellectual argument and meditation." These days, most sleep therapists would give precisely the opposite advice. They would also avoid attaching moral judgment to insomnia in the way Plato does, although they might concede that mental preparation for sleep can make a significant impact on the quality of sleep. But for Plato, there is no such thing as the sleep of reason. The body may rest. But the mind needs to be controlled until dawn. Indeed, the most famous image in *The Republic* is the image of the cave. Plato likens human existence to people living in a dark cavern, looking at shadows of animals and other figures. These shadows are but poor copies of real animals, the ideal forms at whose existence the shadows hint. We may in our lives get to know good people and from this begin to understand what goodness might be. But a good person is only a shadow of an ideal called "the good." For Plato, true knowledge is of ideals such as "the good." The journey of philosophy is from the cave into the first light of dawn.

For Aristotle, on the other hand, philosophy is not the work of dawn but of high noon. It needs the hard light of day. Like Plato, Aristotle had a lot to say about the way human beings should organize themselves into communities. Plato's *Laws* begins with one person asking another whether or not God is the originator of laws. Aristotle's best-known work on the same subject, *Politics*, begins with the line "Our own observation tells us that every state is an association of persons formed with a view to some good purpose." In other words, while Plato starts in the abstract—in the realm of the ideal—Aristotle starts with observation, with experience. One looks at his feet, the other beyond the far horizon.

The result with Aristotle is a colorful melange of sense and nonsense. In his treatise *On Dreams*, he says that the surface of a mirror will always appear red to a woman when she is having her period. This is because her eyes are full of blood at that time. But in the same breath, he suggests evocatively that dreams are caused because the mind in a human body is a bit like a passenger in a vehicle. The mind still keeps moving for a time after the body in which it is traveling has stopped. It has a momentum of its own.

Aristotle's small treatise *On Sleep and Sleeplessness* is likewise so wide-eyed that it is sometimes blind. And yet it also has moments of genuine understanding. It goes into elaborate detail about humors evaporating from food in the stomach, rising through the body because they are hot and entering the brain, having a narcotic effect. This is the philosopher's way of saying that everyone falls asleep after a good lunch. The sleeper nods. Their eyelids feel like a dead weight. They can't stand anymore. Aristotle says this is because the humors entering the head have made the head heavier. People wake when digestion is complete and the heat flows out of the head back down to the body. Thus when they wake, people find their heads are lighter and they can hold them up. He notes that babies sleep a lot because they have large heads in proportion to their bodies. He also says that the bigger a head the more likely someone is to sleep: "Dwarfish or big-headed types are addicted to sleep." Strange to say, modern medicine will find more than a grain of truth in all this, as we will discover later in the night. It will also find truth in his contention that sleep is something peculiar to creatures with hearts and, therefore, with blood. Aristotle notes that melancholics are hard to accommodate in his scheme because they eat a lot but don't sleep much.

But it is the opening and close of Aristotle's treatise that contains its most tantalizing wisdom. Aristotle concludes by saying that we sleep simply because it is in our nature to do so and that no animal can live in defiance of its inherent nature. He leaves up in the air the famous question from which he set out. Sleep could be an activity of the body or the soul or both. For once, he doesn't have an opinion.

MIDNIGHT

[1999]

Midnight is nowhere near the middle of the night; most people are only in the shallows of their sleep when the calendar sweeps out the old day and delivers a new one in its place. But it is still the hour when coaches turn back into pumpkins.

On the night of December 31, 1999, I sat up to watch the end of the world, an event that had been scheduled for midnight that day. For several years, humanity had been on notice about the Y2K bug, a piece of mischief that was going to cause planes to fall out of the sky because the computers that ran our lives had never been told that one day 99 would tick over to 00. I spent the evening with a group of fellow priests who were on holiday; they might otherwise have taken a professional interest in the apocalypse but they weren't too concerned, except for one who wanted to finish his doctorate before it happened. A venerable father took a laconic view of the situation. He told me that the apocalypse would hardly be the end of the world.

This hadn't stopped years of scaremongering, which is always good for the economy. A paper manufacturer had banked on the slogan "Y2K. If in doubt, print it out." Firms appointed Y2K compliance officers and householders stockpiled baked beans and bottled water. In the end, nothing happened. Perhaps we are yet to feel the full impact of the Y2K bug, but so far it's been quiet.

The whole pointless kerfuffle about the Y2K bug is a reminder that both the calendar and the clock are artificial constructs, human fences imposed over the natural landscape of time. The year 2000 for western Christians was 5760 for Jewish people, 2544 for Buddhists, 1716 for Coptics, and 1420 for Muslims. It's odd how religions, which are supposed to be a celebration of timelessness, have been so tangled up in measuring time. Let's not even start on the arguments about the date of Easter.

Sleep, however, does create a rhythm of time. Research has found that when people are deprived of any external prompts such as windows and clocks, they will settle into a pattern of sleep that follows a cycle of twenty-four hours, usually just a little longer. In other words, days and months and years are not inventions. The tools we use to number them are.

The way sleep ebbs and flows over a day is called a circadian rhythm. Jet lag is a problem because it disrupts this rhythm: the traveler's internal clock gets unhinged from the external clock, and common wisdom says that flying east is worse than flying west because it is more difficult to cope with shortening days than lengthening ones. Shift work can produce the same effects as long-distance travel without the pluses: there aren't too many people who send cheery postcards from the lunchroom in the wee hours of the morning.

Different people have different rhythms, and the rhythm changes over the course of a lifetime. There are genuinely such things as larks and owls, people who are more alert in the morning and those who can solve crossword puzzles at midnight. Teenagers tend to be owls. They aren't just refusing to go to bed to annoy their parents. It appears that with all the extra work expected of hormones in adolescence, and all the other secrets that need to be explored, an eighteen-year-old body doesn't get around to secreting melatonin, the hormone produced in the pineal glad to organize sleep and the circadian rhythm, until about 11:00 PM. So when a teenager or young adult turns in to bed at 3:00 AM and stays there until midday, it is physiological rather than rebellious behavior. In later years, melatonin starts to be released about 9:00 PM, the same time that body temperature begins to fall, another prelude to sleep. There has been some sturdy evidence to suggest that teenagers do better in schools that take their circadian rhythms into account and start classes later in

the morning. It is possible that adolescents are less adolescent during the holidays, when they can sleep according to their body clocks. Moodiness and irritability may be signs of poor sleep. Old people produce less melatonin: they tend to become larks, getting up earlier and sleeping less overall, often in broken stretches. The elderly who wake at night worrying about their adult children would most probably wake up anyway. It's good of their children to oblige them with a reason other than advancing years.

Midnight is no more than a line in the air. It is neither the darkest part of the night nor the coldest, nor the time when anyone is most likely to howl at the moon. Yet it is a powerful image in demonology; midnight is the so-called "witches sabbath" when creatures from the dark side do their thing. But Christians also celebrate Midnight Mass on Christmas Eve, a feast whose timing was established to get a free ride on the back of earlier festivals marking the longest night of the year in the Northern Hemisphere. The image of Christmas is the star; it is a festival not of light but of light in darkness, one whose deepest roots are old. Midnight has political nuances: the independence of India at that moment on August 15, 1947, has occasioned books such as *Midnight's Children* and *Freedom at Midnight.* In the sphere of private experience, midnight has also long been held as a pregnant moment, apt for both godly and ungodly visitings. Samuel Taylor Coleridge was a tortured insomniac; his sleep had been land-mined by opium. Yet one of his most tender poems, "Frost at Midnight," written at the close of the 18th century, celebrates a poised moment when all the world stands still. In practice, for Coleridge, such moments were pure fiction. But they are a beautiful fiction nonetheless. Coleridge writes,

> The inmates of my cottage, all at rest,
> Have left me to that solitude, which suits

Abstruser musings: save that at my side
My cradled infant slumbers peacefully.
'Tis calm indeed! So calm, that it disturbs
And vexes meditation with its strange
And extreme silentness.

The tales of *A Thousand and One Nights* start at midnight. They are sometimes called *The Arabian Nights* and sometimes *The Scheherazade*, but really they should be called *A Thousand Nights and One Night*, the extra night on top of the perfect thousand being a gesture toward infinity, a concept enshrined within the Islamic cultures that, in various places over hundreds of years, first told these joyous and wily stories. They include "Sinbad the Sailor" and "Aladdin and his Lamp" and "Ali Baba and the Forty Thieves." They are huddled together under an umbrella narrative that goes like this: There once was a king called Shahryar who was cheated on by his wife, so he decided to take his revenge on all womankind, marrying a virgin every night, having sex, and then putting her to the sword at dawn before she could cheat on him. It is understandable that some readers don't get beyond this point; the text is blasé about these lives. After three years, however, the people are starting to get a bit restless and flee from the city until, in the words of Richard Burton's 19th-century translation, "there remained not in the city a young person fit for carnal copulation." A number of readers who have struggled past the three hundred beheadings find this a bit much: the sex drought seems more of a worry than the deaths. The chief Wazir is grief-stricken because he has run out of virgins to offer the king, and he fears for his life. But the Wazir has two daughters: Shahrazad and Dunyazad. Shahrazad has read a thousand books. She has studied science, philosophy, and poetry. She is wise and witty and has a bright idea. She marries the king, arranging for Dunyazad to be present in the bridal chamber when "the king arose and did away with his bride's maidenhead and the three fell asleep." At midnight, Dunyazad asks her sister to tell a story. Shahrazad obliges, but at dawn when she is due to lose her head, the story is not finished. The King agrees to allow her to finish it the next night. But by then the story has doors opening off it leading into other stories and the king is entranced, ready to be lead through interlocking rooms and corridors of these beguiling stories.

Many of the stories are both profound and profane at the same time. There's one little one about Abu Hasan, a nomad who moves to the city and marries. He becomes a rich man and his wife dies. All this takes about four lines: the stories can be biblical in their ability to leave out the stuff you really want to know and leave in the stuff you don't. Abu Hasan's friends urge him to marry again, but he resists until finally he agrees to marry a woman whose beauty, we are told, is like that of a midnight star reflected in the ocean. So the biggest wedding feast ever seen in that country takes place, and in due course, the steward arrives to accompany Abu Hasan to the chamber where his bride is "displayed in her seven dresses and one more." This is the moment of sexual tension in the story. Abu Hasan arises solemnly from the table, but at that very moment, he farts. Not just any fart, but a fart to do justice to the banquet he has just eaten, a fart that Burton describes as "great and terrible," a fart so loud that every guest "talked aloud and made as though he had heard nothing, fearing for his life." Farting has had a varied relationship with bedtime rituals: in some cultures it is accepted as an economical means for heating up the bed, in others it is grounds for divorce.

Abu Hasan was so humiliated by his fart that he left his own wedding, went down to the port, and got into a boat for India where he stayed for ten years. At the end of that time, he was homesick and longed to hear his mother tongue once again. So he disguised himself as a dervish and crept home by a secret route "enduring a thousand hardships of hunger, thirst and fatigue; and braving a thousand dangers from the lion, the snake and the Ghul." The ghul is a demon that robs graves, often disguised as a hyena. Eventually, however, Abu Hasan reaches his own country, his own town. And there he hears a ten-year-old girl talking to her mother. The girl asks her mother when she was born. The mother replies, "Thou wast born, O my daughter, on the very night when Abu Hasan farted." Abu Hsan realizes that, far from having been forgotten, his disgrace is now the event from which time is measured. So he gets up and goes back to India where he lives out the rest of his days.

For all its silliness, this little story charms its way into a deeper place. Abu Hasan tries to deal with his disgrace first by running and then by disguise. But the past is patient. It waits for him to come back to it. Time stands still for as long as it takes the proud and restless to listen to what it has to say.

Shahrazad weaves a great quilt of stories. A thousand and one midnights come and go, and the tales become even more captivating. During this time, Shahrazad presents the king with three sons, so she did most of her storytelling while pregnant, another poignant image. After so long, the king is a new man, transformed by the power of all he has heard.

The Thousand Nights and One Night is a model of an entire civilization: we all tell stories to save our lives, and we listen to them to save the lives of others. Chaucer's *Canterbury Tales* is a great adventure in which stories are told to pass the time. But Shahrazad tells stories for the opposite reason: to hold time still, to allow one midnight to last forever. The final night ends with a blessing: "Glory be to the living in who dieth not and in whose hands are the keys of the unseen and the seen. Glory be to the one whom the shifts of time waste not away."

12:02AM

[1915, 1916, 1918, 1939, 1943, etc.]

I once worked in an old folks' home where there was a man, Charlie, who took particular care drying his feet. Every morning, I got him out of the bath and sat him on his bed where he started on his big toe with a towel. I then went and had a cup of tea; when I returned Charlie would be just about finished with the little toe on the other foot. Sixty years earlier, Charlie had been a teenager in the trenches on the Western Front.

"I had two long years with wet feet," he said. "You know, it's impossible to sleep with wet feet."

Charlie spent World War I with the watch his late father had given him stopped at 12:02.

"I never lost it. The spring had broken but it reminded me of him. Besides, it doesn't matter what time it is when you can't sleep."

"I never thought about that."

"Besides, I liked to think that it had stopped at the start of a brand new day. I became thankful for every brand new day I survived to see."

He wrote home to his mother and mentioned the feet. It was the least horrific thing he could think to write. She wrote back telling him not to be so silly and to put on dry shoes and socks.

"She had no idea," he said. "Thank God she had no idea."

"She must have been glad you survived."

"She changed the sheets on my bed every fortnight when I was away in case I turned up unannounced. She knew I'd be tired, she could imagine that much. The whole time I was in the army, I used to fantasize about clean sheets and a dry mattress. But when I finally got back to them, I couldn't sleep properly. I used to toss and turn."

"Did this improve?"

"Even now I wake in the night and feel the thud of shells. I never hear them. I feel them. I have nightmares."

Charlie was among the millions of people who feel the impact of trauma on their sleep.

A good number of bedtime stories begin with the words "Once upon a time." I do like that strange expression "upon a time." It makes time seem solid, like a table, something that can support weighty objects, which is what happens when we sit upon a cushion, for example. The Charles Perrault story "Hop O'-My-Thumb" begins "Once upon a time, there was a woodcutter and his wife . . ." Let's think for the moment about the figure of the woodcutter.

The woodcutter provides the perfect entry point to another world. This is the person who goes into the primeval forest, the figure who steps into the dark. The woodcutter brings that darkness into the light of civilization. He (or she, but mostly it's he) turns ancient trees into new timber or even firewood. The woodcutter is a bridge between the ruled and the unruly, between nature and nurture, between control and its opposite. He is a liminal figure, someone who stands at the door, the ideal person for stories that help people set out upon the adventure of sleep. There is a known world on this side of sleep with the comfort of beds and fireplaces. There is a forest on the other side.

Sadly, this picture does not always correspond to reality. The world on the other side of sleep bears a close relationship to what we experience on this side of sleep. Those who experience traumatic lives will, more often than not, have traumatic sleep. And many people who manage, often courageously, to live with past traumas in a constructive way during their waking hours will be ambushed by them at night. For many, the aftermath of trauma is worse at night than at any other time. When there's nothing to distract it and keep it busy, your own mind can be most unpleasant company. The devil makes work for it.

Hansel and Gretel is an example of a story that begins "Once upon a time, near a great forest, there lived a poor woodcutter and his wife and his two children." Note they are *his* children. The wife is a stepmother, the most maligned character in fairy tales. It's a shame that stepmothers all over the world have to bear this injustice. Anyway, the stepmother in this particular story is pretty bad because she wants the kids to die.

Sleep plays no small part in *Hansel and Gretel.* The children can't sleep for hunger, which is how they come to hear of the plan to abandon them to the wild animals in the forest. Hansel waits for the adults to fall asleep before going out to get the pebbles that will guide them home. The adults are able to leave the children in the forest because on both occasions, Hansel and Gretel fall conveniently asleep.

The most traumatic character is the old hag who lives in the gingerbread house. She built the house in order to entice children. She puts Hansel in a cage to fatten him up. Luckily, Gretel is able to push her into the oven, and they escape, taking the old hag's jewels with them. They reach home safely. The wife is dead. The father is overjoyed. They all live happily ever after.

Fairy tales are supposed to insinuate themselves into some of the less readily visited parts of our consciousness. A number, including not just this one but also *Little Red Riding Hood,* include accounts of children being eaten. I have no clue about the psychology of this. One researcher, Bruno Bettelheim, believed (rather implausibly) that such stories were meant to help bedtime listeners get past their oral fixation—that is, get to sleep without sucking on a bottle or a thumb or the edge of a pillow. The mouth, they imply, is a deadly weapon; it's best to sleep with it closed.

Hansel and Gretel is an extraordinary catalogue of child abuse. It includes the death of a mother, starvation, abandonment, attempted murder, theft, imprisonment, and violence. There is even a high-sugar diet to add to that list. "They all lived happily ever after" doesn't seem a credible place to end. Surely it's more reasonable to expect PTSD ever after. To be fair, the brothers Grimm did end with a short paragraph that was omitted from the Golden Treasury that ushered us to sleep as kids. It reads, "My fairy tale is done. See the mouse run. Whoever catches it gets to make a great big fur hat out of it."

In other words, the storyteller breaks the spell and says that the story exists only in a bubble of make-believe.

But actually, this description has the whole thing upside down. *Hansel and Gretel* doesn't describe traumatic situations for the sake of dealing with them. It deals with trauma that existed before the story, and the roots of this story are very old. The key to an appreciation of this is the figure of the old hag or old witch. The old hag lives on in our world in an expression that is part and parcel of any understanding of sleep, especially post-traumatic sleep. That word is *nightmare.* "Mare"— the second part of the word—refers to a hag, or witch, possibly even a demon. There was once a belief that a hag would come during the night and sit on your chest, thus constricting breathing and movement. Those who have experienced sleep paralysis may well consider this to be a better description of what is going on than pages of scientific text. It certainly captures the feeling and even the panic that can come with it. A nightmare was originally a nocturnal visit from a creature like the hag who put Hansel in a cage. Now the word is more commonly used to describe the memory of trauma, not the actual experience of it. But the feeling of a creature sitting on your chest is still an evocative description of what is going on. The neurologist Oliver Sacks said that the supernatural figures of folklore—devils, witches, or hags—can have an important role in helping us at least to describe our experiences: "We make narratives for a nocturnal experience which is common, real, and physiologically based." Sacks believed that the word nightmare should be represented as night-mare, just so we don't lose sight of the image of the old hag.

Nightmares are a common part of the landscape of PTSD. As the human family conceives of an ever-growing list of ways to bring stress into the world, there is more and more need to understand PTSD. It can originate in many experiences: grief, loss, bullying, sexual abuse, violence, injury, and so on. It is even possible to traumatize yourself, as when, for example, you damage a relationship that is important to you. Many people who experience PTSD say that a decent night's sleep is one of the things they most urgently crave and most deeply miss. One expert in the field, professor Kevin Gournay, writes:

> Sleeping problems are an extremely common consequence of PTSD. You need to remember that PTSD causes very high levels of physical arousal, and in such a state it is very difficult for the body to relax and for sleep to take over—even if you are feeling very tired. In addition, sleeping problems can be caused by the intrusive dreams and nightmares that are another common manifestation. Although the symptoms of PTSD may subside over time, sleeping problems often become prominent.

There are thousands, if not tens of thousands, of stories that bear this out. The stories of soldiers returning from war are a painful part of the phenomenon. When J. R. R. Tolkien returned from the Western Front toward the end of 1916, he would have been counted among the lucky ones to have survived at all. He was sent home with trench foot, hardly a pleasant condition, having witnessed the horrors of the Somme. His wife, Edith, complained that he spent almost two years after his return in bed while she struggled to look after a young family and cope with the effects of a difficult childbirth. It was in bed that Middle Earth was conceived. Whether or not Middle Earth is the stuff of nightmare has been discussed by readers. After his long journey, Frodo is damaged; anniversaries become difficult for him. "I am wounded," he says at the end of the saga. "It will never really heal."

In some circumstances, desperation to do something about sleep provides the very weapon that returned soldiers use to kill themselves. James Prascevic served with the Australian Defense Force in Timor, Iran, and Afghanistan. He eventually broke a leg while parachuting; this was the trigger that led to the onset of depression and, from there, to an investigation of the turmoil faced by those who have experienced war at first hand. He writes of the tragic use of one drug, Seroquel, prescribed as a sleep medication. An overdose can cause death. Some have resorted to that extreme.

It is good to know the history of the place where you sleep. Sometimes you stumble across it by accident.

One Christmas a few years ago, when Benedict was eleven and the twins had turned nine, we invited John, an old friend, to join us for lunch. By that point, we were living in a house in a leafy suburb. John was in his eighties and moved slowly with the help of a cane, so I went to pick him up to deliver him to our house.

In the car, John asked where we heading.

I mentioned the name of our suburb.

"Oh, really?" he said. "That's where I grew up."

"What street?" I asked.

"You wouldn't know it. It's just a little street."

It turned out that I did know it: it was our street.

John had grown up a few doors from the house in which we now lived. His family house was still standing. We stopped outside it for a moment in the car. John was silent. Memories seemed to be coming back to him, and he was unsure whether or not to welcome them. John said that his father had been in World War I, starting at Gallipoli in 1915 and remaining in active service until near the end of the war in 1918. He was one of the fortunate few who survived the entire tragedy. He came home and worked hard to qualify for a place in the Taxation Office so that he could support his family. The young John started life on our street in 1929 and went off from there to school where, coincidentally enough, I was now working. He was a member of the class of 1946.

As we arrived at our front veranda, John stopped for a moment and looked down at his feet. "I remember this place," he said. He put another foot forward. "I remember these steps. This was the Franklins' place."

It turned out that Mrs. Franklin had been a great friend of John's mother, so John had spent many days as a boy in the building that was now our house. It stirred memories that had long been dormant.

"Mr. Franklin had been badly affected by the war," he said. "In those days, people just said politely that he couldn't settle to anything after he came back. But everyone knew what it meant. These days, we'd call it post-traumatic stress."

"Were there many returned soldiers here?" I asked.

"Yes," John said. "The whole street."

I learned that all these houses were built by the War Housing Authority and had all been occupied by former soldiers.

One by one, John pointed out places where the inhabitants had been affected by shell shock, including the one opposite us. It was a street in which fragile people struggled to hold down jobs and share the peace they had fought for. It was a place of nightmares and broken sleep. It was a place whose certainties had been deeply challenged.

"The women used to get together in your house to support each other," John told me. "Mrs. Franklin was a kind of focal point, possibly because her husband was one of the worst affected. None of them got much sleep, because they all had husbands who screamed in the night. I can remember my own father screaming. He had terrible nightmares. So the women brought the children here so they could take turns having a nap during the day. That's how they coped. The sleeplessness was like a fog that spread over the whole street."

These stories changed my relationship with my home. It hasn't always been mine, and it won't always be mine. Sometimes, I think about the Franklins and remind myself that there was a lot of experience here before ever I was even born. I sleep in a place that had been used to help draw the poison of war from the wounds of broken soldiers.

There are countless stories of this kind.

Quiet suburbs often throw a blanket over unquiet experiences. One of Australia's most revered military commanders retired to a house very close to ours. He was Harold Elliott, better known as Pompey. Elliott was wounded on the first day of the landing at Gallipoli in 1915 and, like so many others, could easily have died. He became a great general, largely because authority never disguised his humanity. Pompey Elliott was in touch with primal instincts. A lawyer by profession, he was a crusty old so-and-so who was famously rude to everybody, especially his superiors. On the other hand, when Pompey threatened to shoot a man who lit up a cigarette at a dangerous moment in the trenches, the offender's brother, standing nearby, said he'd shoot the general if he did any such thing.

Elliott admired the man for speaking up a

promoted.

Stories like this abound. Ellio

tates of protocol. According to h

so disheveled in appearance that he w

London for impersonating a high-ranking

posed to have their shirts hanging out the wa

letters he wrote home from the front to his two v

beautiful.

Elliott happened to command two battalions at what

good reason, described as the worst day in Australian history.

at Fromelles, his units alone saw more than fifteen hundred men k

a single period of twenty-four hours. Elliott made it clear to those

had set up this disaster that it was a pathetic idea from the outset. The historian Arthur Bazley wrote that "no one who was present will ever forget the picture of him, the tears streaming down his face, as he shook hands with the returning survivors." Elliott was a leader of people, not numbers. He was unafraid to express his grief. A Lt. Schroder described Pompey at Fromelles: "A word for a wounded man here, a pat of approbation to a bleary-eyed digger there, he missed nobody. He never spoke a word all the way back to [headquarters] but went straight inside, put his head in his hands, and sobbed his heart out."

Elliott fought long and hard after the war for recognition of soldiers who suffered invisible wounds, those whose minds were damaged by the experience. Their needs were usually given short shrift. Elliott was appalled by how returned soldiers, because of either physical or mental injury, found it hard to find adequate work in the 1920s. He was devastated that former soldiers suffered worst in the Great Depression. He never recovered from war himself. He endured years of sleepless anxiety. Ross

ites, "He, too, was plagued by nightmares and ghastly flash-
he not only relived front-line horrors he had repeatedly wit-
en more distressing was the memory of all those times he had
iged to order subordinates to undertake perilous assignments."
ot entered the Senate in 1919, but his grief was cyclonic. He took
wn life in Malvern, not far from where we live, in March 1931. Wars
ays come home, even if the soldiers don't. Sleep has a cavernous
emory. It holds on to things that, rationally, we may prefer to let go. That is what a nightmare does: It sits on your chest. It pins you to the very bed on which you are unable to find true rest. It is like the old hag in *Hansel and Gretel* who keeps you in a cage and keeps fattening you up until she eats you alive.

One approach to trauma is called stoicism, a set of practices that is much maligned or at least misunderstood. The clichéd version of stoicism is "Grin and bear it" or "What doesn't kill you makes you stronger." I don't want to disparage these adages because there is an entire self-improvement industry built around every one of them, and it's not my place to deny a living to those who earn their money waging PowerPoint presentations against the world. The problem is that all these sentiments lack compassion and empathy for the people who struggle with life and then, on top of that, are expected to struggle with the fact that they are struggling.

Besides, such sayings don't represent Stoicism. Stoicism is not an unfeeling and aggressive attack on the things life throws in our path. It is a contemplative form of life. It is a way of gently changing the way things are by first accepting that they are. It understands that we, ourselves, are also part of the way things are. Meditation is rightly hailed as an aid to sleep; properly understood, Stoicism is a meditative approach to life. It

does not isolate an activity called meditation. It suggests that mediation is a bit like breathing, something we'd be silly to stop.

Stoics get their name from the Greek word *stoa*, meaning porch, because the original disciples of Zeno, their ancient founder, met on a porch. But *porch* is a word like *woodcutter*. It stands between two worlds: inside and outside. Stoicism does precisely that. It negotiates a path between our inner experience and our outer reality. A classic stoic distinction is between pain and suffering. Pain is part of the objective order. If someone hits you, you will experience pain. But suffering is a subjective response to that pain. The memory of a punch long after the physical wound has healed is suffering. A nightmare that relives the punch is suffering. Bad sleep years after a car crash is suffering. How can we honestly acknowledge pain, which is inevitable, while reducing the residual suffering it leaves?

The stoics were interested in this, and I have a lot of sympathy for the idea of philosophical therapy, the belief that listening to the wisdom of philosophers can shed light on even mundane problems. Philosophical therapy is meditation that is not afraid of an active mind, even a searching mind. The Roman philosophers Cicero and Seneca were Stoics. So, too, was the philosopher emperor Marcus Aurelius, a man whose *Meditations* is still a source of sanity around the world. He was a political leader whose greatest legacy was not political. He was the head of a vast military force whose greatest insights were about surrender. In his collected meditations, he wrote, "Withdraw into yourself. It is the nature of the rational directing mind to be self-content with acting rightly and the calm it thereby enjoys." In another section, he writes about coping with pain: "Unbearable pain carries us off; chronic pain can be borne. The mind preserves its own serenity and the directing reason is not impaired by pain. It is for the parts injured by the pain to protest if they can."

Stoicism puts great faith in our rational abilities, and this is both its strength and weakness. It believes that an active mind is needed to create a still person. The mind is like a surfboard: It is a place to balance in a sea that never sleeps. It creates balance and poise and stillness, but the moment it stops, it topples over.

Marcus Aurelius also thought about sleep. He believed that it is the rational mind that gets us out of bed. He wrote, "When you are reluctant to get up from your sleep, remind yourself that it is your constitution and man's nature to perform social acts, whereas sleep is something you share with dumb animals." He also said that, every time we wake from sleep, the first thing we need to do is actively remind ourselves that we are not going to be buffeted by the opinions of others or the happenings of the day. We will remain at peace because we are one with the whole world, not its victim. He believed that we can shape our own minds such that, no matter what happens, we don't have to be imprisoned by fear and anxiety. This is even true of post-traumatic sleep and nightmares. The important thing is to be in control: "Sober up, recall yourself, shake off your sleep once more: realize they were mere dreams that troubled you, and now that you are awake again look on these things as you would have looked on a dream."

If only it were so easy.

The idea of creating the habits of your mind does have an honorable history. It is something shared across religions; it unites pagan philosophers such as Marcus Aurelius with Christian theologians such as St. Augustine and St. Thomas Aquinas. Nobody reaches peace through passivity; that is achieved through hard labor.

This kind of thinking is certainly still current. Contemporary researchers have suggested that the brain has a kind of neuroplasticity that allows it to reshape itself—or, rather, the owner to reshape it. It is possible that we think of the mind as a train, tied to a certain groove or track, destined to go in a certain direction. In fact, it is more like a boat that can harness the elements, even if sometimes they seem hostile, to navigate a fresh course. The theories of neuroplasticity could have implications for post-traumatic sleep. Revisiting trauma, which seems to descend uninvited on a sleeping person, could be a habit of mind that is possible, with a lot of time and help, to change.

The best-selling writer Norman Doidge, author of *The Brain That Changes Itself*, is a well-known proponent of these beliefs. He says it is possible to work with the plastic nature of the mind to change the way that memories are "re-transcribed" as the physical makeup on the brain is renewed. One of his patients, known as "Mr. L," suffered the loss of his mother when he was twenty-six months old. He was sent away from his siblings to a childless aunt and subsequently had great problems forming bonds in relationships. He had recurrent dreams whose theme was losing something. He turned his mother into a ghost. Doidge worked with Mr. L to turn a nightmare into a real memory of a real mother who had once loved him. Doidge writes,

> Mr. L did not get better all at once. He had first to experience cycles of separations, dreams, depressions and insights—the repetition or "working through" required for long-term change. New ways of relating had to be learned, wiring new neurons together, and old ways of responding had to be unlearned, weakening neuronal links.

Mr. L's ghost needed to be replaced by reality. Doidge continues:

> Why are dreams so important in analysis, and what is their relationship to plastic change? Patients are often haunted by

recurring dreams of their traumas and awaken in terror. As long as they remain ill, these dreams don't change their basic structure. The neural network that represents the trauma—such as Mr. L's dream that he was missing something—is persistently reactivated, without being re-transcribed. Should these traumatized patients get better, these nightmares gradually become less frightening.

Sometimes people discover creative ways to find sleep at the end of a dark tunnel.

Ray Parkin was a prisoner of war on the notorious Thai-Burma railway during World War II. Even before he got there, he'd been through hell and high water. After having enlisted in the navy as a teenager in 1928, he survived the sinking of his ship, the HMAS *Perth* in March 1942. He was then part of a small group adrift in a dingy. He was finally captured and held by the Japanese on Java. None of this was pleasant. But hard labor on the railway trumped all these horrors.

Parkin was also an artist and many of the paintings and drawings he completed while in captivity are the most enduring depiction we have of the ordeal endured by him ands his comrades. Some of these images survived because he hid them in the makeshift operating table of his close friend, the legendary doctor "Weary" Dunlop. One of the things that helped Parkin get through this period was that he resolutely kept doing drawings of beautiful things, such as flowers. He made a decision to see more than darkness. He was especially fascinated by the butterflies that abounded in the area, creatures that seemed lighter than air. Wounded men would look at his drawings as a form of therapy. Even the guards seemed uplifted by them.

Needless to say, trauma affected the very sleep the men needed to cope with illnesses that ranged from malaria to dysentery to cholera, not to mention the extraordinary work demands of building the railways and the brutality of the captors. In his diaries, Parkin maked brief observations such as this: "Had a pretty bad night. But it was full of thought." He was ever-optimistic despite the circumtances.

Parkin pinned the bodies of butterflies to the roof of the makeshift building in which sick men tried to recuperate. Later he told his biographer, "The Sistine Chapel had nothing on it." Parkin gave the men something beautiful to contemplate as they tried to find sleep. The wings moved gently in the breeze. The world beneath those butterflies was a more reassuring place.

1:30 AM

[2006]

When our three children were little, we found that we could get them to sleep in the car, so we spent a lot of time there. On Friday nights, we would line them up in their three car seats and go on a date. We'd drive to the nearest Asian food outlet. By the time we'd done the thirty miles, they had nodded off. We bought noodles and parked outside a café where we could look in through the window and imagine we were part of an adult world. Then we got a takeaway coffee and drove to a spot where we could overlook the Hume Highway and tell each other escape stories. Before long, we checked the little ones in the rearview mirror and got talking about them, wishing they would grow up but still be little.

Once we got as far as the coast, where we met a family from Western Australia who lived in a rough and ready mobile home. They spent their lives doing laps of the continent, crawling from beach to beach, picking up odd jobs to cover the basics. The twins in that family had just turned ten.

"We never set out to be wanderers," explained the mother. "I guess it just happened."

"It didn't just happen," said the dad. "Nothing just happens."

You meet philosophers everywhere.

The mother explained that when their twins were little, the only place they could get them to sleep was in the car. So they started doing longer and longer trips, switching off driving so that one would drive while the other got some shut-eye in the passenger seat.

Over time, this had become a lifestyle. Ten years later, they were still driving around. They had done countless thousand miles.

"We've been back to Perth a few times, but I can't sleep at home anymore," said Dad. "So we've rented the house. That's how we pay the bills.

We teach the kids ourselves. If they need to know about something, we drive there. It's better than looking it up on the Net."

About that time, a Cambodian woman turned up in town with her two sons. It appeared that some kind of arrangement had been brokered with one of the older bachelors in town, and suddenly these three people arrived looking as though they had landed on another planet, unable to understand ordinary English let alone the granite dialect used in our district. The newcomers spent a lot of time indoors. The mother found work on a chicken farm an hour away packing the eggs laid by battery hens, bringing home cartons of chipped ones that could be eaten but not sold. She started at 5:00 AM and had to leave home in the early hours to get there. The stepfather took little interest in the boys; as a matter of fact, he didn't take much interest in anything and locals were surprised that he had heard of Cambodia let alone managed to recruit a partner from there. Each day, the elder boy caught the bus over to the local high school where there was no kind of support for someone in his predicament. The poor kid was spending most of the day sitting in the corridor. When the bus returned in the afternoon, his eyes would be fixed straight in front, his head turned to stone. The younger boy seemed to cope better; he was still at an age where you can catch a language a bit like catching a cold.

Their neighbor Tony was appalled by the older boy's suffering and decided to do something. He did some research and discovered that a school in Canberra had a program for students who struggled with English. But there was no bus to get there. So Tony organized a car pool, approaching various people who went to the city for work. At the time, I was doing some work at the university, so a couple of afternoons a week I'd pick up the young man in our old red car with a clock on the dashboard that had died at 1:30 AM ages before. The boy greeted me with a curt nod, pulled his seat belt around himself like a zen master pulling on a robe and, within one hundred meters, was fast asleep, sitting perfectly upright. He never moved for the whole journey, his head grazing the low roof of the car. When we got home an hour later, the dashboard clock said it was still 1:30. Time had stood still, just like in the fairy tales. The boy always woke at the same point as we came into town. He said good-bye with a small nod and disappeared into the stepfather's house. Tony explained that the young man had spent time in a monastery when he was a boy and his disciplined posture was a legacy of this. But the ability to use sleep to control reality was a skill he had developed all on his own. I wondered what kind of trauma his refugee experience brought with it.

At the end of the year, his mother appeared in our driveway with two cartons of eggs.

"I thank you," she said solemnly. "I thank you for my boy."

None of those eggs were chipped.

Three years later, the boy was studying architecture at a university in another state. He was a different character now, all because Tony went out of his way to make sure there were cars for him to sleep in.

As much as it is true that trauma can disrupt sleep, there are circumstances in which certain trauma can induce excessive sleep, almost as though the mind were closing down to protect itself. I thought this was happening to the boy. It can certainly happen to prisoners. There have been stories in the media of asylum seekers, who have been held indefinitely in off-shore detention centers by the Australian government, sleeping for fifteen hours a day, an image of hopelessness and despair. There is a fascinating documentary called *Chasing Asylum* made by filmmaker Eva Orner. It uses footage that has been surreptitiously obtained from these off-shore processing centers on Manus Island and Nauru; many of the characters need to remain anonymous. The film shows people with nothing to do but sleep. They sleep to the point of self-extinction.

I once met a Chinese priest, Archbishop Dominic Tang, who had spent twenty-two years in jail (in my opinion, for no good reason), much of it in solitary confinement. He was less that five feet tall and exuded a deep gentleness, but he was made of tough stuff. When asked how he'd coped, he said he put himself onto the monastic regime he'd learned as a novice so that he would have a different task, usually spiritual, to perform every hour or half hour. He made a structure for his day. He washed and shaved slowly, deliberately, consciously.

"Otherwise sleep too much," he said in broken English.

1:50 AM

[2000]

Experts have counted two hundred or more references to sleep in the work of William Shakespeare. This level of interest is striking until you learn that Shakespeare and his wife, Anne Hathaway, had twins: a boy and a girl, Judith and Hamnet, who were born in February 1585, twenty-one months after the birth of their first child, Susanna. (Tragically, Hamnet died at the tender age of just eleven.) The presence of twins surely put sleep high on Shakespeare's agenda, not to mention the extraordinary pressure of time under which he worked both day and night. On the other hand, maybe Shakespeare took such a vigorous interest in sleep simply because he took a vigorous interest in just about everything.

It's hard to escape the conclusion that Shakespeare's slumbering Falstaff, described in *Henry IV Part 1* as a "bed-presser" and a "huge hill of flesh" suffers from sleep apnea, aggravated by grog. When Peto finds him "fast asleep behind the arras and snorting like a horse," Prince Hal remarks, "Hark how hard he fetches breath." Shakespeare describes him with textbook accuracy.

On the other hand, Julius Caesar asks to be surrounded by men "that are fat" and "such as sleep at nights." In this case, Shakespeare was mistaken if he thought fat people are better, as opposed to longer, sleepers. They only look that way, as anyone with sleep apnea will tell you. But if he was saying that better sleepers make better leaders, he may well have had a point. Brutus, one of the traitors, does not sleep on the night before he joins the assassination.

Sleep is a character in a number of Shakespeare's plays and poems. In *The Tempest*, Prospero says:

> We are such stuff
> As dreams are made of, and our little life
> Is rounded with a sleep.

His "Sonnet XXVII" also begins with a nod to nocturnal routines:

> Weary with toil, I haste me to my bed,
> The dear repose for limbs with travel tired;
> But then begins a journey in my head
> To work my mind, when body's work's expired.

In *Macbeth*, there's also a fine speech in praise of sleep, delivered at the very moment when Macbeth kills the king:

> Sleep that knits up the ravelled sleave of care,
> The death of each day's life, sore labour's bath,
> Balm of hurt minds, great nature's second course,
> Chief nourisher in life's feast.

It is curious that Macbeth delivers these thoughts before he even tries to wipe the blood off his hands, but sleep was obviously something that weighed on Shakespeare's mind. When Macbeth starts to fall apart, his wife doesn't attribute his poor state of mind to the bloodbath he has created in the pursuit of power. She simply tells him, "You lack the season of all natures, sleep."

Whether it was caused by parenting-induced exhaustion or not, Shakespeare had a fascination with sleep that never left him. He even made specific mention in his will, bequeathing to his wife Anne "the second-best bed with the furniture." (Most opinions concur that "the furniture" means the bed linen; the idea of a second-best bed is more confusing. Scholars have been divided between those who regard the second-best bed as second-best—and hence claim the will was a slight to Anne—and those who regard it, for various recondite reasons, as really the best. Among their arguments is the claim that the best bed was the one kept for guests but this was really second-best, because any couple with a brain kept the best bed for themselves but had visitors believe it was second-best.)

A bed is a bit like a family dining table. It bears many stories. Yet some cultures and some people are coy about beds, perhaps because of their association with sex. In the famous comedy *I Love Lucy*, Lucy and her husband, Ricky, always had separate beds. If Ricky was in bed, he'd be reading the paper with his pajamas ironed and buttoned up to his neck; if Lucy was in bed, she'd be fully robed and wearing lipstick. That was considered proper.

But a bed is not just a piece of furniture; people seldom feel as intimately attached to, say, a couch as they do to a bed. For most of the story of sleep, the efforts of bed builders have gone into appearances rather than anything else; this is strange because a bed does its best work in the dark. But beds have needed to fulfill both public and private purposes, and where these have been in tension, the public seem to have taken precedence. When Tutankhamun's tomb was opened in 1922, the discoverers were impressed by his range of beds, made from precious materials such as ivory, ebony, and gold and featuring elaborate carvings of cats, included more for their sacredness than their sleepiness. Comfort isn't such an important factor for the dead, whose backs are as bad as they're going to get. While alive, Henry VIII—a big bloke—had a bed that was over three square meters. (He did have six wives but not all at once. The bed was more about making a statement to the underlings.) Meanwhile, Louis XIV of France had 413 beds.

Of course, no matter what other purposes they might serve, beds still serve their primary service as places to sleep. The expression "sleep tight" comes from a practice, developed in restoration England, of putting a mattress on a lattice of ropes that would then be pulled tight to provide both comfort and firmness. The method also worked to discourage unwanted visitors, such as rodents. It is from this custom that we get the saying "Good night, sleep tight. Don't let the bedbugs bite."

Besides the threats of insomnia and bad dreams, bedbugs are perhaps the worst nightime visitor. The common bedbug, *Cimex lectularius*, is a resilient little bugger. A typical bedbug measures less than five millimeters across and is flat, making it apt to fit in the tiniest places. It wouldn't be such a nuisance if it just ate waste, as the demonstrator implied, but it's a little Dracula, sucking blood by night, injecting victims with both an anticoagulant and an anesthetic when it does so. This means that you don't feel the dreadful itch until well after the burglar has come and gone, often leaving the tap still running so that you find blood spots on the mattress. Bedbugs are patient. They can remain dormant for eighteen months between feeds. Despite popular myth, they work across barriers of income and class. The introduction of steel beds led to a decrease in their prevalence, as did cotton sheets because they could be boiled. But pest control experts are at a loss to explain why bedbugs have become so numerous once again over the last few years. They breed like bugs.

Many people share a bed, but everybody sleeps alone. Sleep, like death, is a door you can only go through on your own.

During the final week in which I was working as a priest, I was called to a suburban motel in the middle of the morning because a woman had woken that day to find her husband dead in the bed beside her. The pair had been married for fifty years and never slept apart. They had traveled together and, every year, made a pilgrimage to the city for the Spring Racing Carnival where they always stayed in the same room at the same inexpensive motel. It was a second home to them, so much so that once they had found the previous year's racing slip under the bed, where it had remained undisturbed for twelve months.

They were a devoted, if eccentric, couple. The receptionist who greeted me said that the woman was coping with remarkable composure. Around 8:00 AM, she had rung the front office and asked if it were too late to change the order for two breakfasts to one breakfast.

"May I ask why?" inquired the receptionist.

"Oh, it's just that my husband is dead in the bed," the woman apparently replied.

A doctor was called, who duly pronounced that the husband must have passed away about half past one.

"I don't think so," said the woman. "I was watching TV then. I would have noticed."

Then they called me, the priest, and I was asked to sit a few minutes with the body while the widow slipped into the little bathroom to apply her makeup.

"Thanks for waiting," she said. "I want to look my best. You can put the TV on if you want, although there's not much on at the moment, I don't think. I was channel surfing before you arrived."

Twenty minutes later she emerged and we were just getting ready to start prayers when somebody arrived to service the room.

"Would you mind coming back to make the bed later?" asked the widow. "I'm afraid my husband is still in bed."

"No problem," the maid replied. "I'll come back later."

"Thank you," said the widow, adding, "We may need fresh sheets."

We knelt by the bedside, and as we prayed, the woman reached across and held the hand of her dead husband, a gesture I found comforting despite the fact that it wasn't supposed to be me who needed comfort. It was the first sign that the body in the bed was anything more to her than a stage prop. Later, I tried to suggest to her that she might be in shock because someone so familiar to her—a person who had shared her bed for fifty years—had just slipped away without so much as ruffling the sheets.

"That," she replied, "is the sign of a really good mattress."

Even for the living, sleep can be a profound form of absence. That's why Santa and the tooth fairy and the sandman all come at night; they know that your body is in the bed but that you yourself have slipped out for a moment, so they can do their thing.

Some years ago, when I was still in the order, I received a call from the police station. A homeless man had wandered into the station and handed my wallet across the counter. I thanked the officer but told him that this was impossible as my wallet was in my beside drawer where I had put it the night before.

At the time, I was living in a terrace house belonging to a religious community. My bedroom was upstairs and at the back; access was provided by a flight and a half of creaky stairs, and the door was so tight in the jam that often it needed a shoulder to persuade it to let anyone in. The room was so small that everyone assumed it had once been servants' quarters and that the servants had been a chaste breed: a single narrow bed all but filled the space, leaving room only for a wardrobe at the foot of the bed and a small table on the other side of the bed from the door. To get to the table, you had to shuffle sideways like a crab around the bed and past the wardrobe.

"I'm sorry, officer," I said, "but it simply can't be my wallet." I didn't go into the salubrious details of religious accommodation. "It must belong to someone else," I said.

However, just to be sure, I went upstairs to check. My keys were in the drawer where I had put them. And my sunglasses were in the place where I emptied my pockets every night. But the wallet was gone. I rang the officer back.

"I think I need to come and see this wallet," I told him.

It turns out that the man had come into my room in the early hours of the night, slipped around the bed, and pilfered the wallet—all while I was asleep. I never noticed a thing.

"Where were you all this time?" asked the officer.

"I was in bed," I told him. "Asleep."

"You may as well have been anywhere," he said.

"I suppose."

I retrieved my wallet and went back home. When telling one of my brethren about what had happened, I marveled at the fact that the man had known to come into my room.

"He probably just followed the snoring," the priest said.

2:00 AM

[1856]

Florence Nightingale spent most of her long life in bed. This is not the way she is usually remembered. Her bed became a stage on which she performed her austere eccentricity.

In March 1855, a man named Alexis Soyer arrived to assist Nightingale at the Barrack Hospital in Scutari (now Üsküdar), a small town across the strait from Constantinople. It was here that casualties were being ferried, across the Black Sea from battles fought in places such as Balaclava, Sevastopol, and Inkerman. Soyer, a Frenchman, had come to the Crimean War at his own expense to help an English woman. Soyer was a man with a taste for the finer things and had made his name as the chef at London's Reform Club, a high-class venue, where he had worked since its opening in 1837. (The wager that sends Jules Verne's Phileas Fogg and his valet Passepartout around the world in eighty days was laid at the Reform Club; J. M. Barrie, creator of *Peter Pan*, was later to be elected a member.)

Following the death of his wife, Elizabeth, in 1842, Soyer had taken himself to Ireland to feed the hungry. There he had invented a kind of soup kitchen suited to the needs of the destitute. He soon wanted the whole world to sit at his table; his books had visionary titles such as *Soyer's Charitable Cookery* and *A Shilling Cookery for the People.* On the way to the Crimea, Soyer stopped off in Marseilles in order to inveigle a recipe for bouillabaisse out of a rival kitchen, but it was not a recipe he was going to be able to use immediately. The sick and wounded soldiers for whom he had decided he was going to cook at Scutari had, in the previous twelve months, been close to starvation. The only reason most of them didn't die of hunger was that disease got them first.

Soyer had learned about what was going on in the Crimea in the same way that the rest of England found out: he read the paper. The invention of the telegraph meant that news of military disaster and administrative ineptitude now reached the taxpaying public with what was, for the authorities at least, uncomfortable speed. The previous October, correspondents for the *Times* (London) such as William Russell and Thomas Chenery had alerted the public to the gross neglect of the injured, most of whom were "left to expire in agony." Chenery commented that "the commonest appliances of a work-house sick ward are wanting, and that the men must die through the medical staff of the British army having forgotten that old rags are necessary for the dressing of wounds." The hospital was portrayed as a morgue. Readers were outraged, and their reaction, both in its immediacy and intensity, marked a shift in public sensibility: they were beginning to think of ordinary soldiers less as beasts of burden, with few interests beyond sex and sleep, and more as heroes, or at least figures of sympathy. Unfortunately, their commanders were slow to pick up on this change; most of them had more feeling for livestock than for their men. Sidney Herbert, the Minister at War, decided that the only person who could do something about the appalling conditions was his friend, Florence Nightingale, who had infuriated her well-heeled family some years earlier by taking up nursing.

Nightingale arrived at Scutari in November 1854, not long after the Charge of the Light Brigade and within a month of the most alarming reports appearing in the *Times*. She brought with her a team of almost forty women, including a contingent of nuns. The women found they were not welcome; they were given five small rooms in which to sleep, one of which still contained the body of a dead Russian general whose hair had entirely fallen out.

In the twenty-one months that followed, Florence Nightingale was to become—perhaps after Queen Victoria—the most famous woman in the world. She is remembered as the woman who stayed up late, looking after sick and wounded soldiers for whom the experience of the hospital had become more fearful than battle itself. After 150 years, the image of the "lady with the lamp" is about all that remains of the Crimean War, a pointless military escapade that has been all but wiped from memory.

The war came about in the first place for no better reason than because England and France wanted to wag their finger at Russia and be nice to Turkey. England hadn't had a decent war for years and was worried about its military machine getting rusty. It wanted to show off the fact that its navy could move an army to the other side of Europe within a couple of weeks. Indeed, it could. But once it got to the shores of what is now Ukraine, it didn't have a clue what to do next.

Upon the arrival of the English, the Russians made soothing noises, but the English decided that it would be a shame to have come all this way and go home without a decent outing. The Battle for Sevastopol ensued, and that city's streets were soon churned into mud. The whole situation was absurd. The injured were tied onto donkeys, manhandled onto ships, and then shipped to the hospital in Scutari, where disease was rife, sanitation was laughable, and the limited supply of water was poisonous. The soldiers would have fared better if they'd stayed behind and fended for themselves in the mud.

English soldiers were known for their colorful turns of phrase, but in Scutari, a pall of silence hung over the place. When Nightingale suggested to her superiors that a screen might be erected to prevent soldiers having to watch others undergoing amputations, her request was met with incomprehension. The same thing happened when she suggested that the soldiers might drink less if they were actually able to send some of their pay home. When she argued that a teacher might be employed to educate the men as they recovered, the boss told her not to "spoil the brutes." She was up against the rule book. It didn't matter how many people died as long as correct military procedure was followed. The Crimean War was so mad that it made Nightingale herself look sane. This is saying something. She was no simple sister.

By the time Alexis Soyer arrived on the scene five months after Nightingale, she was beginning to work a transformation. Soyer's flamboyant personality belied his own capacity to get things done—something that endeared him to Nightingale, who, all her life, had been hamstrung by an inability to suffer fools. Nightingale was not a fool. But her soul was made of stainless steel. She was appreciative when Soyer came up with new designs for teapots and camp ovens that could handle the demands of the situation; he developed decent soups and stews. On top of this, Soyer was a writer who sent his own material back to the papers and who, when he got home, published a book with another fine title: *Soyer's Culinary Campaign.* In it, he describes walking the wards in the early hours of the morning and encountering the figure of Florence Nightingale. His depiction of her at 2:00 AM has become part of the iconography of nursing:

> As we turned the angle of the long corridor to the right, we perceived, at a great distance, a faint light flying from bed to bed, like a will-o'-the wisp flickering in a meadow on a summer's eve, which at last rested upon one spot. . . .
>
> But alas! As we approached we perceived our mistake. A group in the shape of a silhouette unfolded its outline in light shade. As we came nearer and nearer the picture burst upon us. A dying soldier was half reclining upon his bed. Life, you could observe, was fast bidding him adieu; Death, that implacable deity, was anxiously waiting for his soul to convey it to its eternal destination.

> But stop! Near him was a guardian angel, sitting at the foot of his bed, and most devotedly engaged in pencilling down his last wishes to be despatched to his homely friends and relations. A watch and a few more trinkets were consigned to the care of the writer; a lighted lamp was held by another person, and threw a painful yellow *coloris* over that mournful picture, which a Rembrandt alone could have traced, but which everybody, as long as the world lasts, would have understood, felt and admired. It was then near two o'clock in the morning.

Another contemporary report, first published in the *Times*, is even better known: "When all the medical officers have retired for the night and silence and darkness have settled down upon these miles of prostrate sick, she may be observed alone, with a little lamp in her hand, making her solitary round."

The lady with the lamp was a woman of unbending religious conviction. She worshipped an unreasonable God, one who infuriated her by not doing as she said. Sometime in the 1870s, she wrote on a piece of scrap paper: "I MUST remember that God is not my private secretary." Nightingale was essentially a Platonist, a believer that this world serves only to provide fuzzy clues about an ideal world that exists beyond it. It is no wonder that one of the few strong friendships she formed in old age was with Benjamin Jowett, the celebrated translator of Plato. For Nightingale, the world was essentially a dark place, a cave of shadows, and humankind was, in a spiritual sense, nocturnal. As far as she was concerned, all of us fumble across a dark terrain, helped by the light of a moon that sometimes emerges from behind clouds to throw the shadows into relief and even to create an impression of fragile beauty. The light of day was never going to be part of this world, althoug now and again, a stronger illumination might cause trouble. This had been her experience when, on February 7, 1837, at the age of seventeen, she received her calling. "God spoke to me and called me to his service," she later claimed. That call, she believed, meant she was never to marry, despite at least two tempting proposals, and it was so significant that she celebrated its anniversary throughout her life, always with a sense of duty rather than joy. She had been called in the same year that Victoria came to the throne, and as the queen was celebrating her golden jubilee in 1887, Nightingale was marking hers with an unflinching spirit.

Nightingale grew to love her patients and came to think of herself as a virgin mother, a higher order of motherhood in her mind than the unvirgin type. Yet at the same time, the hospital ward was a place of shades, part of the obstacle course this world presents to those who want to qualify for the real world. It was part of the penumbra that reality cast upon the earth, and she, too, moved across it like a shade. One of the soldiers who was able to write sent a letter home, saying, "We lay there by hundreds; but we could kiss her shadow as it fell and lay our heads on the pillow again content."

In the days before electric light, Nightingale's lamp cast a long shadow over the beds, up the walls, and along the ceiling. Her presence was like that. The men worshipped her.

Florence Nightingale knew the world had a lesson to learn from the Crimean War, and she was determined to teach it. One of the messages she wanted to get across was about the importance of bedding. It seems obvious now, but she was among the first to wonder if it was such a good idea for a wounded soldier to be wrapped in a bloody blanket, packed off in it to hospital, and then, in all likelihood, buried in it a short time later.

In *Notes on Nursing*, the influential guidebook she published in 1860, she brings her customary obsession with detail to bear on the nature of bedding: "Feverishness is generally supposed to be a symptom of fever—in nine cases out of ten it is a symptom of bedding."

Above all else, Nightingale believed that bedrooms need to be well aired. She was an evangelist for the power of fresh air. Likewise, bedclothes need to be aired and dry. Bedrooms need good light and, if possible, a view. Beds should be low to the floor so that a patient doesn't feel the roof is closing in upon him and burying him alive. Nobody should ever put their head under the blankets. The role of the pillow is to take the weight off the chest so that the patient can breathe more easily. Careless nurses fold blankets back and leave too much weight on the chest of the patient. Nothing should ever be damp. Florence Nightingale was a force against moisture in all its forms:

> My heart always sinks within me when I hear the good housewife, of every class, say, "I assure you the bed has been well slept in," and one can only hope it is not true. What? Is the bed already saturated with somebody else's damp before my patient comes to exhale into it his own damp? Has it not had a single chance to be aired? No, not one. "It has been slept in every night."

In *Notes on Nursing*, Nightingale describes patients as "entirely, or almost entirely, prisoners to bed." By the time she wrote these words, she was herself in bed. Nightingale had left the Crimean War in mid-1856, at the age of thirty-six. She died in 1910 at the age of ninety. In those fifty-four years, she made no public appearances, gave no interviews, and did all in her power to extinguish her celebrity. The lady with the lamp became an even more shadowy figure. She spent almost all her time in bed. Her bed became a place of refuge, if not safety: there was one occasion in which, lying in bed, a cistern fell onto her through the ceiling.

There are umpteen possible reasons why Florence Nightingale took to her bed and stayed there for more than half a century. No single item on the following list can explain what was going on, and some of these suggestions may be mutually exclusive, although Florence Nightingale was no stranger to paradox.

First, it's quite possible that she was simply tired. One theory is that she suffered from chronic fatigue syndrome (CFS) and indeed her birthday, May 12, has become international CFS awareness day. There are a lot of myths about CFS, sometimes known as myalgic encephalomyelitis (ME). People who suffer from this mysterious malady don't necessarily sleep all day; it is more common for ME to play its baffling hand in the disturbed, inadequate, or shallow sleep of those forced to share their lives with it.

It's also quite possible that she had another illness. Various possiblities have been suggested. One biographer endorses the theory that Nightingale suffered from brucellosis, a bacterial infection sometimes called Malta fever. The disease's long list of symptoms includes depression, poor sleep, exhaustion, and episodic paralysis. And there are some possible hints of this in Nightingale's writings. In 1875, she wrote to a friend, "A more dreadful thing than being cut short by death is being cut short by life in a paralysed state." (The Irish poet W. B. Yeats was diagnosed with Malta fever in Italy in the late 1920s and he, too, found himself glued to bed, sleeping long hours. When awake, Yeats coped by drinking champagne and reading pulp fiction, remedies unavailable to the austere Nightingale.)

Perhaps Nightingale suffered from a mental issue that caused her to take to her bed. She may have had post-traumatic stress disorder. It's hard to imagine more traumatic conditions than those she found at Scutari. But it's also quite possible that she suffered from agoraphobia, a fear of open spaces, that had struck her later in life, also conceivably the product of trauma.

Or it may be that the reason is even simpler than that. What if she simply wanted to avoid her family? Florence was always impatient with women, perhaps because they were not as easily directed as men, and much of her frustration was reserved for her mother, Fanny, and sister, Parthenope, known as Parthe. Nightingale believed that the two women dined out on her fame when it suited them but never lifted a finger to help in her campaigns for better sanitation. The relationship was dysfunctional. By staying in bed, she could shut the door on them. When her mother proposed visiting her in 1865, Nightingale wrote back, "Even ten minutes' talk with those I love best secures me a night of agony and a week of feverish exhaustion." She said that her mother could enter her room at 4:00 PM and kiss her, but under no circumstances was she to say anything.

Nightingale certainly was particular about her schedule and visitors, but it's also important to remember that other people have done the same thing (although perhaps not quite as thoroughly). In 1745, at the age of nineteen, a man named Josiah Carlton took to his bed and stayed there until he died in 1805. Carlton said that he wanted to avoid sinning. Charles Darwin also took to his bed for an extended period of time after he returned from five vigorous years on the *Beagle*. No one knew what was wrong with him, but he was so mysteriously and so completely indisposed that he was scarcely able to get out for his daughter's wedding. John Milton, Mark Twain, Robert Louis Stevenson, and Winston Churchill all set up offices in their beds. In 1909, G. K. Chesterton published an essay called "On Lying in Bed," in which he remarks that he'd be happy to remain on his back all day "if only one had a coloured pencil long enough to draw on the ceiling." His advice: "If you do lie in bed, be sure you do it without any reason or justification at all."

The difference between Nightingale and most of the great bed workers of history is that the latter have managed to face the world by early afternoon. One notable exception was Cardinal Richelieu who, as chief minister of France from 1624 to 1642, was not to be inconvenienced by rising from bed. He simply had the bed carried to wherever he needed to do business, even if walls had to be removed to get him in. He gave his own twist to the idea of pillow diplomacy, the ancient art of sleeping in high places, often to achieve low ends. Brian Wilson, founder of the Beach Boys, also took to his bed for long periods, hardly the image of sunshine, surf, and outdoor fun so seriously projected by that band, although perhaps it was a wave bed.

What is true is that, even from the depths of her bed, Nightingale was nothing if not prolific. From under her covers, Nightingale started a training school for nurses. She designed hospitals. She wrote endless submissions, briefs, and textbooks. She ran a Royal Commission from bed. She organized sanitation for the whole of India, a place she never visited. She advised on medical requirements for the Civil War in America, another place she never got to. Indeed, as her world became increasingly abstract she acquired encyclopedic knowledge of all sorts of places she never set foot. In all, over fourteen thousand of her letters survive, most of them including the time of day at which they were written, often 4:00 AM or 5:00 AM. She was simply too busy to get out of bed.

Nightingale may have been busy and engaged, but in many ways, her bed provided a buffer—not just from her family but also from the public at large. Maybe she believed that if she was going to be the author of change she needed to be anonymous. When she left the Crimea, Florence Nightingale returned to England as "Miss Smith," deliberately avoiding the heroine's welcome that was in store for her. She made a conscious judgment that the changes she wanted to bring about in public health could never be brought about by grandstanding. She orchestrated inquiries and a Royal Commission from behind closed doors. Over the years, the mere rumor of her presence in public could trigger near-hysteria. She knew that applause meant the end of a performance, so she never took the bow.

Being in bed also seemed to be a way for Nightingale to transcend her body. In her youth, she had castigated herself for the habit of "dreaming," by which she meant absenting herself from reality. The real problem for her was that she actually enjoyed "dreaming"; those dreams may have even had an element of sexual fantasy. Her frenetic approach to doing good was at least partly a way of dealing with this. Her virtue was compulsive. She had a deep-seated fear of disorder or chaos. In Crimea, she had confronted the darkest of these forces and put them in their place in about twenty-one months. Yet fifty years were nowhere near enough to get herself into the same kind of order. Her theology increasingly became an attempt to force God around to her way of thinking, a strenuous effort to control the uncontrollable. She wrote a kind of anxious spirituality, keen to find a system for God that would spare her an experience of God. In the year she published *Notes on Nursing*, she also wrote *Suggestions for Thought to the Searchers after Truth among the Artizans of England*, a work whose title took up most of the first of three volumes. In *Eminent Victorians*, Lytton Strachey memorably unfolds Nightingale's cold idea of a God for whom the tragedies in the Crimea were a necessary part of an eternal plan:

> Yet her conception of God was certainly not orthodox. She felt towards Him as she might have felt towards a glorified sanitary

> engineer; and in some of her speculations she seems hardly to distinguish between the Deity and the Drains. As one turns over these singular pages, one has the impression that Miss Nightingale has got the Almighty too into her clutches and that, if He is not careful, she will kill Him with overwork.

In 1864, Nightingale wrote "Note on the Aboriginal Races of Australia" to be delivered as a paper at the annual meeting of the National Association for the Promotion of Social Science, held in York that year. Heaven only knows how and why she got involved in this discussion, although her attitude shows her signature sharpness. She was never afraid to have a point of view:

> In their present state, very few of the human race are lower in the scale of civilisation than these poor people: excepting indeed those who trample upon and oppress them—who introduce among them the vices of European (so-called) civilisation.

What is more interesting is the characteristic way in which she sorts out the situation of indigenous Australians according to her own religious agenda, one in which God is needed to provide a clear sense of duty. Her religion was a prophylactic against loss of control. It was all about work, not rest:

> In dealing with uncivilised races it has hitherto been too often the case that the Roman Catholic Missionary has believed: "sprinkle this child with Holy water and then the sooner it dies the better" and that the Protestant Missionary has believed "make this child capable of understanding the truths of religion and then our

work is done." But the wiser Missionary of this day says: "What is the use of reading and writing to the natives—it does not give him a living. Show him his duty to God. And teach him how to plough. Otherwise, he does but fall into vice worse than before. Ceres comes before Minerva.

Yet, all her efforts did not succeed in putting the world in order. There was a rough sea that kept lapping around the feet of her bed. In the mid 1870s, Nightingale was lying in bed, watching the shapes cast by her nightlight. She got to thinking about Scutari and sat up to write herself a note: "Am I she who once stood on that Crimean height? 'The lady with the lamp shall stand.' The lamp shows me only my utter shipwreck."

In the year before Florence Nightingale was born, John Keats wrote his wonderful "Ode to a Nightingale," a poem in which the bird is an elusive symbol of immortality. Like a number of the so-called romantic poets, Keats had a fascination with sleep, an experience that provided a handy gray area between consciousness and unconsciousness, between reality and unreality. Another example is Wordsworth's "A slumber did my spirit seal," a beguiling short poem that struggles with both mortality and immortality, and also Coleridge's *The Rime of the Ancient Mariner* in which the storyteller gains a moment's respite from his story:

> Oh sleep! it is a gentle thing,
> Beloved from pole to pole!
> To Mary Queen the praise be given!
> She sent the gentle sleep from Heaven,
> That slid into my soul.

Keats's "Ode to a Nightingale" ends with a confused return from reverie to reality: "Was it a vision, or a waking dream? / Fled is that music:— Do I wake or sleep?"

Florence Nightingale asked herself the same question: Was this life a form of waking or sleeping? Her worst fear was to resemble Keats's bird that "wast not born for death." When friends told her that she had become immortal, that countless thousands of young girls had been named after her, she couldn't react. On return from the Crimean War, she fully expected to die within the next year or two. It was a source of confusion to her that she didn't and that she lived on and on and on. Like the voice in Keats's poem, she was "half in love with easeful Death."

The nightingale is the opposite of the lark; it sings at night. Yet a better emblem of Florence Nightingale is Shakespeare's "obscure bird," the night bird that doesn't sing. One of the closest relationships Florence formed was with an owl called Athena. She found Athena on a visit to the Parthenon in Greece in 1850, a period in her young life in which she was badly sleep-deprived and in which she was obsessed by her proclivity for dreaming. "I had no wish on earth but to sleep," she wrote. "I lay in bed and called on God to save me." So, at the home of Western philosophy, God sent her an owl, which slept in her pocket and soon ate her pet cicada, whom she had named Plato. Athena, named after the goddess of the philosopher's city who is often represented by an owl, died the week before Nightingale left for Crimea; Florence blamed her sister Parthe for killing Athena. Soon after, as Florence was beginning to come to grips with the enormity of the task at Scutari, Parthe sent her an illustrated biography of the poor bird, which she had written. When the soldiers discovered Nightingale's secret grief, they tried to find her another owl.

Legend has it that the owl is too wise to sleep. It is, like Florence

Nightingale, wide-eyed. When Keats wrote "To Sleep," he prayed for respite from his own mind:

> Upon my pillow, breeding many woes,—
> Save me from curious Conscience that still lords
> Its strength for darkness, burrowing like the mole;
> Turn the key deftly in the oiled wards,
> And seal the hushed Casket of my Soul.

The cliché-makers like to tell us that bed should be kept for sleep and sex and nothing else. They have a point. There is a discipline called sleep hygiene that teaches that we can improve our sleep by having a less-cluttered sleep space, that a bedroom should be furnished with a kind of sacred minimalism. So no TV in the bedroom, no laptops or briefcases. No eating in bed either. The bed is an altar, and sleep needs to be attended by certain safe rituals, all of which move the body and soul to a point of surrender, not least by being familiar, unchanging, and, therefore, comforting. What the sleep sanitation people forget to mention is that sleep and sex are cousins. This is why a place that is bad for one tends to be bad for the other as well. But the two activities have a deeper affinity: you reach both of them over the lip of a cliff, and you have to let go in order to fall into either. Both sleep and sex are places where the ego can lose its stranglehold on your identity—experiences in which, if you are lucky, you can hand over the keys to your own being.

Florence Nightingale achieved more in bed than the vast majority of people achieve out of it. She spent years in bed not sleeping and saved thousands of lives in the process; both her energy and her lack of it could be staggering. There is plenty owing to this nightingale. At the same time, she is a bundle of contradictions. She believed that God called her to anonymous and thankless service, yet she kept fierce control of her life, narrowing it down to an area of six feet by three in order that she could save the whole world. Her problem was that the lady with the lamp could never escape her own shadow.

2:06AM

[2007]

As he approached his fourth birthday, we decided it was time for our Benedict to start making his own bed. All we wanted him to do was to pull up his blankets and put his pajamas under his pillow. We soon lightened up on the second requirement, as the pillow was needed to shelter all the little toys that Benny took with him on the journey to the Land of Nod, so there wasn't much room left for pajamas. We concentrated instead on getting him to arrange the blankets neatly. He was resistant to the idea.

"If God made the world," he asked, "why can't he make my bed?"

It's not a bad question. I wish I knew the answer.

"God made the world so you'd have a place to make your bed," I ventured, not really sure of my theological footing but relieved that Benny was at least prepared to think about what I'd said.

"So why does Mummy make your bed?" he asked.

Bed-making is one of our culture's more curious forms of behavior. I suppose that just as we need reassuring rituals to get us into bed, so too do we need them to get us out of it again. Making the bed is a way of putting the world in order at the start of another day. It is a comforting work of fiction. In a few minutes, you can impose order and regularity on the tangle of sheets and blankets that represent the third of your life over which you have least control, the hours you've spent in the Land of Nod.

I thought that perhaps our Benny needed the firm hand of Florence Nightingale, a demon for making beds. She was the mother of that fine innovation in the deployment of bed linen, the hospital corner. She wrote, "A true nurse will always make her patient's bed carefully herself."

But a four-year-old is too young for Nightingale. And so we simply tried to explain that a new day was a wonderful thing containing lots to look forward to and making your bed was a way of getting ready for it.

That night, Benny appeared between us in our bed. I looked across to the clock radio. It was 2:06 AM.

"Benny," I said as mildly as possible, "it's nice to see you, but you have your own bed so why don't you go back there."

"I can't," he said. "I've already made it."

Not long afterward, we read about the International Exhibition of Inventions, in Geneva, where an Italian engineer called Enrico Berruti exhibited a machine that he had been working on for ages. It was a bed that made itself, including straightening the duvet and tightening the undersheet. Berruti had devoted years to saving a few minutes. We thought he'd get along fine with our Benny.

In the history of understanding sleep, there is a tension between control and surrender. Is sleep an orderly experience, something reflected in the image of a neatly made bed? Or is it the opposite, a daily encounter with a mysterious realm beyond our consciousness? The demands of life, for most people, seem to be ever increasing. We put more and more effort into staying in control. As always, the way we live affects the way we sleep and the way we think about sleep.

2:10 AM

[1728]

It was Jonathan Swift, the author of *Gulliver's Travels*, who first referred to sleep as "the Land of Nod." Our children loved *Gulliver's Travels*, which had also been a bedtime favorite of mine.

On the whole, it is an invigorating experience when you first encounter a story as a child and then discover it all over again as an adult, possibly even at different stages of adulthood, each time stretching your imagination to reach new places. It's like learning to play in the safety of shallow water before you gain enough understanding and respect of the forces around you to move into the deep. We all pay an enormous price when people first encounter their sacred texts, such as the Bible or the Qur'an, as children but then never learn to read them as adults. Sacred texts need deep water; in the hands of fundamentalists, they are like whales stranded on a beach, thrashing about dangerously, gasping for air.

In my own childhood bedtime reading, *Gulliver's Travels* is paired in my memory with *Robinson Crusoe*, the original version of which was written by Daniel Defoe. Defoe was a prototype of what came to be known as a journalist. He also had sidelines as a spy and an insurance salesman; his eye was as sharp as Swift's, but Defoe was not so much appalled by the human condition as Swift was, as ready to make a buck out of it. He did humanity the disservice of taking it far too seriously.

Robinson Crusoe, written in 1719, and *Gulliver's Travels*, written in 1728, have a few things in common. They were written within the space of a few years of each other and are both fantasies of exploration and discovery. They both use the kind of narratives that reached their apotheosis in the reality TV series *Survivor*. In addition, they both also have extraordinarily long titles, enough to put any librarian off their lunch. Swift's book is called *Travels into Several Remote Nations of the World. In Four*

Parts by Lemuel Gulliver, First a Surgeon, then a Captain of Several Ships. Not to be outdone, Defoe's is called *The Life and Strange Surprizing Adventures of Robinson Crusoe, Of York, Mariner: Who lived Eight and Twenty Years, all alone in an un-inhabited Island on the Coast of America, near the Mouth of the Great River of Oroonoque; Having been cast on Shore by Shipwreck, wherein all the Men perished but himself. With An Account how he was at last as strangely deliver'd by Pyrates.* It sounds more like the title of a bill before Congress than an adventure story. But despite their similarities, *Gulliver's Travels* and *Robinson Crusoe* are born from radically different understandings of the world. And different views of humanity can be seen in different attitudes to sleep.

When the eponymous Robinson Crusoe finally comes ashore on his castaway island, he immediately starts putting his new world in order. He realizes that he is defenseless, and with undaunted logic, he establishes a safe place for his bed in "a thick bushy tree." Defoe writes,

> And having drank, and put a little tobacco into my mouth to prevent hunger, I went to the tree, and getting up into it, endeavoured to place myself so that if I should sleep I might not fall. And having cut me a short stick, like a truncheon, for my defense, I took up my lodging; and having been excessively fatigued, I fell fast asleep, and slept as comfortably as, I believe, few could have done in my condition, and found myself more refreshed with it than, I think, I ever was on such an occasion.

Crusoe is the embodiment of what used to be called Protestant work ethic, a sturdy and, in many ways, admirable approach to life. It is evident, for example, in the practice of putting clocks on church towers. Religion is supposed to be about timelessness, but when there's a clock on

the steeple, there is a caveat: yes, we believe in eternity, which is why the steeple points so high, but here below you still have to keep an eye on the time and use it prudently. Crusoe often talks about God and the way the Almighty has looked after him and so on, but the reader becomes increasingly skeptical about this. In point of fact, he is a self-made man. His first bed typifies his whole mind-set. He uses reason to rig up safe bedding in a hostile world and is rewarded with sound sleep. As the book goes on, we find that Crusoe can do anything by the combined power of reason and hard work, even overcome mental illness. He comments, "my reason began now to master my despondency." Thousands of people might wish it were so easy. He is never really alone on his island because he is a one-man civilization. The whole point of the novel is that a rational creature "not bred to any trade" has the capacity to re-create the entire apparatus of European civilization. Crusoe does this without a flicker of self-doubt.

Crusoe is obsessed with time and its correct measurement. For him, sleep is not just a personal experience, designed for the well-being of the individual. It is also how we measure the passage of days, and these build into weeks, months, and years—and ultimately into what we call history. Every child knows how many nights it is till Christmas or their birthday. This aspect of sleep is far more important to Crusoe than any of its restorative qualities; indolence is a sin and sleep can be a Trojan horse bringing sloth and laziness into the world. The measurement of time is instrumental in bringing order to existence. *Robinson Crusoe* (if I may use the abbreviated title) is replete with the counting of days, the numbering of history. Crusoe keeps the Sabbath religiously (I am not sure how else it could be kept) and is distraught when he thinks it is possible he has been observing it on the wrong day; he is also punctilious about his diary. More than anything, when he eventually finds a companion, he calls the man

Friday because that is the day on which he found him. Friday is named after a measurement of time. If he'd been found one sleep earlier, he would have been called Thursday.

In *Gulliver's Travels*, when Lemuel Gulliver comes ashore in Lilliput, he, too, is ready for sleep. But here it is a completely different story.

The children's version of *Gulliver's Travels* tends to include Lilliput (where people were small) and Brobdingnag (where they were big) but leaves out the parts that are complex and dark, meaning most of Jonathan Swift's original work. *Gulliver's Travels* is not exactly lighthearted. It is funny, sometimes hilarious, but the humor tears strips off the follies of the human animal. In the bedtime version, the little Lilliputians were cute, and when he reaches Brobdingnag, to the giants it is Gulliver who becomes like the Lillputians and also seems cute. This was far from Swift's intention. The idea of cuteness had no place in his view of the world, nor did ideas such as "sweet" or "delightful." Swift was one of the most savage satirists ever to have put poison in their ink. Just consider his short piece *A Modest Proposal*, which suggests ending famine in Ireland by eating children. Its logic and structure are so impeccable that they end up calling into doubt the functions of logic and structure themselves; this is precisely what Swift intended. He creates an unarguable case that is patently absurd. His use of reason leads to crazy results. Lucky for the pretentious of the world that he did not live past 1745, having endured several years of dementia, a fate he always dreaded. He is buried in St. Patrick's Cathedral in Dublin, an establishment of which he had been the dean. He wrote his own epitaph, which reads angry indignation can no longer lacerate his heart. Swift is one of those who only find rest in death.

Swift was born in 1667, a year after the Great Fire of London, a calamity that may well have been the result of poor sleep. It is said to have begun in a bakery in Pudding Lane just after midnight on September 2 that year, caused by sparks that escaped the fire. (The phrase "just after midnight" is the clue. It suggests that some poor soul had been left to keep an eye on the oven but had nodded off.) However, there were plenty of folk who wanted to blame the fire on terrorists. England was engaged in a war at the time and spies were thought to have crept up the Thames and set fires. The likely real cause was more mundane and less newsworthy—just a simple accident in the bakery.

The Great Fire destroyed thousands and thousands of houses and almost ninety churches, including a great and labyrinthine cathedral. It may have been one of the factors that contributed to Swift's sense of the fragility of civilization. For him, civilization was brittle; the human ego, in contrast, was virtually unbreakable.

There is a great fire in *Gulliver's Travels*, one that threatens the palace of Lilliput. Gulliver stirs from sleep and pisses on it to put it out. This is no ordinary piss. Swift likes us to know that it is the first piss of morning, that pungent and powerful concoction that your body brews while you're asleep. In Gulliver's case, it is all the more offensive because he has been drinking the night before. The palace is saved, but the king is offended because pissing within the royal precinct is illegal.

Swift was both fascinated and appalled by bodily functions. He offers a counterweight to our contemporary culture that wants to believe in a physically sanitized version of the bedroom. This means crisp white sheets, perfumed pajamas, fluffy pillows, and even fresh flowers on the table. The entire nightwear industry is a denial about what really happens during sleep. It's better to think of the old, old joke about the old, old gentleman who goes to the doctor with a list of ailments. The doctor says that he will need a urine sample, a blood sample, a stool sample, and, yes, even a semen sample. The old, old man is hard of hearing.

"What did he say?" he asks his wife.

"Just give him your pajamas," she says.

We like fictional versions of bedrooms, much as we do with bathrooms. But the reality of sleep is nightly sweat, shedding hair, flaking skin, saliva, dry throat, unconscious scratching, dribble, rumbling stomachs, frequent flatulence, strange utterances, various kinds of nocturnal emissions, and occasional incontinence. It's no surprise that some marriages need separate bedrooms to survive. Swift made play with all this, never more so than in the final paragraphs of *Gulliver's Travels*, after the voyager has returned from living as an animal, a brute Yahoo, among the rational horses, the Houyhnhnms. The Houyhnhmns are disgusted that any creature such as Gulliver should wear clothes to bed. When Gulliver returns to England, he is far from happy at being reunited with his wife and family. Five years later, he is still appalled by the smell of humans and revolted by the thought that he was once guilty of "copulating with one of the Yahoo species," the evidence of which is the existence of children. Swift did not much like children. Gulliver buys two horses and prefers to sleep with these. He prefers their smell. "They live in great amity with me," he says; it's the closest Gulliver gets to love.

The poet Peter Steele writes of Swift's fascinated revulsion from anything resembling "ordure" that "it may be that the dream of cleanliness is like the dream of reason . . . he believes in the end that filth is no joke, except in the dark and complicated sense that man himself is a joke." The phrase "dream of reason" is especially apt, both for Swift's time and ours. Swift belonged to an age in thrall to the idea of reason; reason was the celebrity of the 18th century. It has its limitations, but it's surely better than the Kardashians. Swift is one of the few to see reason as a dream. This doesn't mean it is not real; it just means that you need sleep to get there.

All this brings us to Gulliver's first arrival in Lilliput. He comes ashore, walks inland and abandons himself to slumber:

> I was extremely tired, and with that, and the heat of the weather, and about half a pint of brandy that I drank as I left the ship, I found myself much inclined to sleep. I lay down on the grass, which was very short and soft, where I slept sounder than ever I remembered to have done in my life, and, as I reckoned, about nine hours.

Gulliver doesn't create a bed; he just lies down. When he wakes up, as most people will remember, he is tied to the ground with hundreds of "slender ligatures" that are wrapped around his whole body "from my arm-pits to my thighs." It is a wonder that none of this has woken Gulliver, especially having so many ropes, however slender, strapped around his thighs. Gulliver has quite a few resemblances to his creator. Looking at pictures of Swift, with or without his wig, it is clear that he was a jowly individual with a broad neck and a double chin. He was not adverse to wine. He never shared a room with his partner, so much so that people were never entirely sure if the pair were married. I'd wager good money that he was a snorer. Indeed, the evidence all points to a significant case of sleep apnea. If Gulliver is created in Swift's image (as he is in so many ways including his preposterous views on education), it's feasible that Gulliver could have slept through his imprisonment by the Lilliputians.

The contrast between Crusoe and Gulliver is not simply between creating a structure for sleep on the one hand and reckless surrender of consciousness on the other. Crusoe's sleep is part of the order of the world; Gulliver's is a doorway to another world. When he wakes, he is twelve times the size of everyone else. All sense of normal proportion is gone. *Gulliver's Travels* makes endless comedy out of its sense of lost order, lost proportion, and lost perspective. The floating island of Laputa, for example, can move from one time zone to another. It can also move over the cities of its enemies and turn day into night, spoiling the growth of crops. Crusoe could no more play with these ideas than jump over the moon.

Sleep is another country. Swift even gave it a separate name: The Land of Nod. But Swift didn't actually invent the phrase. It actually comes from the first few pages of the Bible. After Cain kills his brother Abel, Cain is forced to become a "fugitive and a vagabond." Like his parents, Adam and Eve, he is forced into exile as a result of his own ego. By the end of the first two generations of the human story, the whole saga is already a mess, characterized by intergenerational family dysfunction. Genesis 4:16 reads, "And Cain went out from the presence of the Lord and dwelt in the Land of Nod, on the East of Eden." The Qur'an tells this story in almost the same way but is more humane and compassionate than Genesis. The Qur'an says mildly that Cain "ended up remorseful" and, as a result, God says that anyone who kills a person either out of revenge or even just "to prevent corruption of the earth," then "it is as if he killed the whole of mankind." Conversely, when anyone saves a life, it is as if they have "saved the whole of mankind." These words are central to the whole texture of the Qur'an.

The Bible seems to be harsher, but its bark can be worse than its bite. The former chief rabbi of the United Hebrew Congregation of the Commonwealth, Jonathan Sacks, has described the Book of Genesis as "the story of human relationship . . . the necessary prelude to Exodus, the story of nations and political systems." He points to the endless stories of conflicts between brothers in Genesis and draws the conclusion that its main theme is "the rejection of rejection." Exile is not death. There is no eye for an eye or tooth for a tooth. Cain killed his brother and he will carry the mark of that forever. But he also gets a fresh start in a place called The Land of Nod. It is to be a new homeland beyond the burdens of waking reality.

Jonathan Swift took the phrase and gave it the meaning with which it is currently associated, namely as a description of sleep. It appears in *Swift's Complete Collection of Genteel and Ingenious Conversation, according to the most polite mode and method now used at Court, and in the best Companies of England*—a story that is neither genteel nor ingenious nor polite. Nor, thank God, is it even complete. It is a desultory display of banalities exchanged between stuffy stereotypes. This explains why no one other than scholars has heard of it. By the end, even the characters in it are falling asleep. It is, admittedly, after two in the morning, and they have been drinking and playing cards. Before he dozes off, The Colonel announces, "I am for the land of Nod." Mr. Neverout replies, "Faith, I'm for Bedfordshire."

This was a significant moment in the history of sleep. It describes sleep as a place of exile, away from family, success, failure, self-image, and the weight of daily life. It is the place to which inadequate humans can escape. It is beyond our borders. It is the daily refugee to which we all turn; we all seek asylum there.

2:15AM

[2007BC]

The Qur'an, the Bible, and the Hebrew scriptures all describe God as the one who "shall neither slumber nor sleep." So if God sometimes appears moody, irascible, unreasonable, and even petulant, lack of sleep may be the explanation. It may just as well explain why God is prepared to put up with so much: an eternity of sleeplessness is enough to lower anyone's resistance.

God is a creature of the night. On page one of the Bible, we are told that God made the night first and then made the day afterward. That has to say something about priorities. The first biblical creation story relates that God's ultimate achievement was not the creation of the firmament of heaven, nor of the fruit tree yielding fruit after his kind, nor of the great whales, nor of everything that creepeth upon the earth, and not even the creation of humankind "in the image of God." The grand finale in God's big production was what happened on the seventh day: He rested. The Qur'an doesn't like this idea, cutting off the creation saga after six days and insisting that God was not the least bit wearied by the whole escapade. But the Bible doesn't say God was tired. On the contrary, it implies that God had a fresh idea and, as a result, simply rested. Indeed, tiredness can be the enemy of rest; we live in a world that is often too tired to sleep.

In the Genesis story, the creation of rest is God's crowning glory, and the ability—for one day a week—to be godlike in taking it easy became the hallmark of the culture that wrote up the story. The capacity to rest without actually sleeping is the most significant accomplishment of the species at the top of the evolutionary ladder. Mosquitoes can't do it. Think of that the next time one torments you at night. You may not believe it at the time, but you are a superior being to that buzzing insect. Otherwise you wouldn't be lying awake at a quarter past two in the morning.

Creation starts with a formless void called chaos and ends with an orderly void called rest. Countless people in every time and culture have found the same thing for themselves: creativity is about creating space, about having the courage to stop, even if it means feeling dizzy, while the world keeps turning. This is the strange essence of the contemplative life, the heart of all great spiritual traditions. The monastery is not a place of escape, quite the opposite. There isn't much you can do to encounter God. You've got a much better chance by not doing. Florence Nightingale never understood this. Her God kept her so busy that the two of them never got to meet.

The irony is that having put rest at the summit of all creation, the Bible story shows God as the great disturber of rest. God gets into people's sleep but not to snuggle up next to them. On page two, in a second creation story, God makes the man fall into a deep sleep so that he can pinch one of his ribs and make him a mate. Abraham, the first of God's great travelers, encounters God in his sleep and the experience is one of terror. It was in a dream by night that God bestowed on Solomon, who'd been caught out sleeping with other gods, the gift of "a wise and understanding heart." One of Job's friends likewise rubs up against God in his sleep and finds "in thoughts from the visions of the night, when deep sleep falleth on men, fear came upon me, and trembling, which made all my bones to shake." An angel appears to Joseph in his sleep and tells him that Mary, the girl to whom he is engaged, is pregnant even though she hasn't had sex but it's okay because the father is the Holy Spirit. This is not the kind of news you want to hear in the cold light of day. Once the child is born, an angel tells him to grab the mother and boy and flee for dear life into the desert. Then later another angel tells him it's safe to go home. It is hardly surprising that legend has it that Joseph died young. He probably stayed up late, dreading that if he went to bed he'd get another of a dream (which is what you experience) and its latent content (which is what it really means). Partly to protect ourselves, we dream in code and the deciphering of those codes is by no means a simple task. Later visitors to Freud's vast oeuvre have scratched their heads about this. How difficult could it be to point out that a pen in a dream might represent a penis? A sword, on the other hand, could represent a penis. So might a rifle. Then again, a shovel might very well represent a penis. Same for a plough. As you advance to another level, you get to ponder dreams that are actually about penises. In these, a penis is a subtle symbol of a penis. Most things seem to represent either a penis or something a penis can fit into. It would be interesting to know just what Freud thought a simple dream might entail. To be fair to him, his range of interpretation was not really so narrow. Dreams could also be about castration.

Despite the obvious disparities between Aristotle and Freud, the two thinkers share an underlying conviction. This is the belief that dreams are the work of the individual dreamer and reflect something of that individual's experience. Most people these days would go along with that in some way, shape, or form.

But it could be that dreams are about the future in a way neither man considered. Aristotle believed that infants did not dream. After all, they have little experience and therefore little to dream about. But these days, as we've seen, we know that newborn children spend about half of their time asleep in REM, a great deal more than at any other part of their lives, and that fetuses spend even more. The evidence all suggests that unborn children spend a lot of time dreaming. But what on earth could they be dreaming about? It's not as if they have issues with life that need processing: sexuality that wants to slip its moorings or bosses that remind them of what their parents used to be like. The answer is complicated

visit in his sleep. The Bible is peppered with dry dreams; dry dreams are the ones in which the dreamer discovers their impotence.

There are almost as many theories about dreams as there are people to dream them up. At the risk of being simplistic, the theories tend to fall into two main groups. The first is that the dreamer has the dream; the second is that the dream has the dreamer. In the first, the dream sheds light on the dreamer by providing clues about the workings of their mind or soul or psyche or whatever you care to call your really private parts. In the second, the dreamer is left with a riddle to solve; the dreamer has to shed light on the dream.

Under the first of these umbrellas, some theories, probably most theories, hold that a dream is a reflection of something going on in the life of its host, whether that something is psychological or purely physical, whether that something is meaningful or whether a dream is just the trash icon on our mental screens, the place to which we drag our psychic junk so that we can delete it from memory. Regardless of the dream's significance or lack of it, a dream is the work of the dreamer; no one else can dream your dreams for you. One of the best-known proponents of this view, at the end of the spectrum that sees dreams as meaningful clues to the waking world, is Sigmund Freud, who wrote in *The Interpretation of Dreams* that "every dream will reveal itself as a psychological structure, full of significance, and one which may be assigned to a specific place in the psychic activities of the waking state." In other words, someone's dreams will take you deep inside the inner workings of that person. Freud builds on the work of Aristotle.

Aristotle did his own thinking in a time and place that had numerous gods to do your thinking on your behalf. In the Greek view of the world, Morpheus, the god responsible for dreams, was the son of Hypnos, who was responsible for sleep and who was, in turn, the son of Nyx, goddess of night. (Incidentally, Nyx had two sons. The other one was Thanatos, death.)

Aristotle believed in putting the ruler over every aspect of human experience and didn’t have much confidence in ideas that were beyond concrete evidence, such as that the gods would be telling us stuff in our pajamas. His writing on dreams is characteristically meticulous, ploughing the field of his inquiry with a fine-tooth comb. He always speaks of “seeing” a dream, not “having” a dream, and he points out that if dreams were vehicles for communication from the gods about the future, then animals wouldn’t dream because why would divine beings be bothered to communicate with lower-order beings. It’s a mystery how Aristotle knew that animals had dreams, but it turns out that he was probably right: more recent studies have shown that most large mammals, and other species as well, have REM sleep, so in all likelihood they dream. They also snore—dogs especially. Aristotle held that the cat curled up in front of the heater is an unworthy receptacle for divine revelation. There are owners who would beg to differ. There’s no point asking dogs about their dreams because sleeping dogs lie.

Aristotle had more to say on the subject of where dreams come from. “If it were God who sent them they would appear by day also,” he said, “and to the wise.” Aristotle was a bit of a snob: for him, the fact that ordinary people happen to have dreams means that, despite appearances, the dreams themselves must be ordinary. “It is absurd to hold that it is God who sends such dreams and yet that he sends them not to the best and the wisest but to any chance persons.” Aristotle thought that dreams were caused by the mind continuing to move after the body has stopped, just as water continues to slop around after the container it is in has come to rest. For him, dreams are like projectiles: they keep going after the thrower

has stopped: “It is the mental picture which arises from the
sense-impressions when one is asleep, in so far as this conc
that is a dream.”

Aristotle believed that dreams are fragile: they are comprise
stimuli that continue all day but that, like the stars, we are onl
notice at night when competing stimuli have toned down. Dreams
reflections on the surface of water: the moment the still water is
they vanish. The minute we move, dreams go the same way. Even af
this time, Aristotle’s theories have a fair bit going for them. But wh
comes to the fact that, after the event, reality sometimes takes the sh
of something that appears to be foretold in a dream, he has a single-wo
explanation: coincidence. Such coincidences may appear frequent, bu
that’s because there are so many dreams taking place in the world on any night of the week that sooner or later one of them is bound to resemble the future. If dreams are a revelation of the divine, then they are only in so far as the whole of nature is a revelation of the divine because dreams are part of nature, not outside it.

Freud applauded Aristotle for not being bullied by a culture that was infatuated by the idea of prophecy. Freud also believed dreams were not about the future; for him they concerned the past. He used them to take him a long way into knowledge of a person; whereas Aristotle was far more inclined than Freud to believe that dreams were a mirage, an illusion, a random collection of mental bricolage. They were not to be trusted. For Freud, on the other hand, dreams are truthful in a way that the conscious mind has learned over the course of a lifetime not to be. In *The Interpretation of Dreams*, he writes, “I must insist that the dream actually does possess a meaning, and that a scientific method of dream-interpretation is possible.” The scientific method to which Freud refers was necessary, because in his view, there was a difference between the manifest content

but probably has a lot to do with the way a fresh brain needs to constantly test its wiring and activate its circuitry. Dreaming in the unborn and in infants may not be visual in the way adult dreams usually are. There are theories, admittedly difficult to test, that relate fetal sleep to the creation of memory and language skills, to the development of cognitive capacity. There is even a hypothesis that adults don't really need REM sleep and therefore don't really need to dream; the process is just a leftover habit from the initial stages of our lives when we needed a special start program to boot up our brains. Dreams are toys we have never let go of because, a lot of the time, they are fun, even if the fun is sometimes a bit like a ride on the ghost train: either you have a good dream that you enjoy or a bad one from which you enjoy waking up.

If it is the case that dreams are of most practical benefit to little people, dreaming may have less to do with dealing with personal experience after it happens and much more to do with creating the capacity to actually have personal experiences later in life. After all, without language and memory, there isn't much prospect of that happening. In these theories, the dream does not explain the dreamer. Rather the needs of the dreamer explain the dream, and all significant dreaming is to the same end—namely, setting up those mysterious qualities of mind that are the basis of those equally strange things called personality and, beyond that, character. These are what make us different from one another. The creation of human individuality and uniqueness gets a big helping hand from the fact we all have the same dreams.

Dreams are one of the more baffling aspects of the Bible. Perhaps the storytellers, whoever they may have been, found in them a handy plot device that allows God to step on stage without being seen. But this explanation is too easy. Biblical dreams are eerie and disquieting stories that go a long way toward capturing the elusive nature of the divine. They sit uncomfortably on that line that divides the ego from something much greater. Generally, biblical dreams tend to shape the dreamer rather than the other way round. Biblical sleepers dream as children but are expected to respond as adults, an irony that sets up some of the poignant tensions in the story. Maybe the dreams are just a way of saying that sometimes, in order to see properly, you have to close your eyes.

Jenny and I named all our children after contemplatives, people whose most potent activity was expressed in stillness. Our little Clare got her name from the woman who was both an anchor and a goad for Francis of Assisi. The medievals tell a great story about Clare and Francis having a meal together; they made their table on the bare ground. Their love was so strong that they never touched the food but spent the time in perfect stillness, totally present to each other. There was so much heat in the room that bystanders thought the house was going to catch fire.

Our Clare's twin, Jacob, got his name from the deepest sleeper in the Bible. There is plenty not to like about the biblical Jacob, a feature he shares with any number of the main characters in the Book. The original Jacob is a figure of gritty determination. His name means "heel" because he was born gripping the heel of his twin brother, Esau, trying to get out before him and so be the firstborn and heir of his father Isaac (whose own name means "laughter," because his mother laughed when strangers told her, at the age of ninety, that she was to have a son; it was a sour laugh that one). The rivalry between Jacob and Esau is painful reading for any parent of twins. Isaac loved Esau but their mother, Rebekah, preferred Jacob. The story goes that Esau became a skillful hunter, but Jacob was a "plain man" who preferred to stay at home. One day, Jacob happened to have made some soup just as Esau walked in from the fields faint with hunger.

Esau was so famished that he exchanged his birthright for a bowl of lentil soup, which remains one of history's most lopsided deals. Jacob didn't hesitate to exploit the vulnerability of his brother. Then, to make it all official, Rebekah came up with a scheme to disguise Jacob in animal skins (to mimic the more hirsute Esau) and have him slip into the presence of his father, who was now almost blind. In this way, Jacob steals the blessing from Isaac that is intended for Esau. When Esau realizes what has happened, it looks like he is going to kill Jacob in revenge, so Rebekah packs Jacob off to her brother, Laban, who lives in the north in Haran. There Jacob falls in love with Laban's daughter, Rachel, and works for his uncle for seven years to secure her hand in marriage. When at long last the time comes for the wedding, Jacob is himself the victim of a trick—and ends up hitched to the elder daughter, Leah. So he knuckles down for another seven years to get Rachel as well. The Bible's first family makes most other families look functional.

All together, Jacob has twelve sons and one daughter by four different women. Jacob's favorite son is Joseph, a preference usually attributed to the fact that, after another long wait, Joseph was the first child he had with Rachel. Jacob buys Joseph a special coat, which provokes jealousy among his siblings, and so the next generation of bad behavior gets under way. But there may well be another reason that Jacob feels something special for Joseph. Joseph understands dreams. Like his father, he knows that however complex life can get there is no guarantee that sleep is going to be any more simple. In the Qur'an also, Joseph is presented as a man at home in the landscape of dreams; the prophet Muhammad was, on occasion, guided by dreams and understood them as part of God's revelation. Indeed, the holy book describes the Night of Destiny, Laylat al-Qadr, which occurred on Mount Hira around 610. On this night, God's word came to the prophet as a source of peace and light in darkness; the description has a dream-like quality. Yet the prophet was also suspicious of the practice of divining dreams, which was part and parcel of the colorful religious climate in which he lived and which he sought to simplify.

The biblical Jacob's sleep took him to places he would rather not have gone. For a bloke who only ever wanted to stay at home, he spent a long time away from it. And it was while sleeping by the side of roads that he had two famous dreams. In the first, which takes place on the way to Haran, he sees a ladder reaching from heaven to Earth with angels going up and down it; God appears, identifies himself by name, and declares that Jacob will inherit the land on which he is sleeping. God promises to be with him and keep him safe. Yet when Jacob wakes up, his memory is one of fear and dread. "Surely God is in this place," he says. He takes the stone that he had used for a pillow, pours oil over it, and makes it a shrine, saying that if God is as good as his word, the pillow will be God's house. This is a tantalizing image of sharing a resting place with God.

Twenty years later, Jacob, now a wealthy man with many children, is returning home from Haran to encounter Esau for the first time since he fled, when he has another dream, again by the side of the road, again when he is sleeping alone. But this dream, less well-known to the popular imagination than the story of Jacob's ladder, is radically different from the one he hosted as a younger man. This time, a mysterious figure wrestles with him until first light and, despite Jacob's pleading, refuses to divulge his name. Rather, the figure changes Jacob's name to Israel, because he was able to hold his own against God. Jacob is wounded in the hip and leaves the encounter with a limp that will slow him down for the rest of his days. Yet he is grateful to have seen God face-to-face and to have got away at all. The reader can't even be sure if it was a dream or something else. It never says that Jacob slept; only that he struggled "until the breaking of day."

There is a lot of growth between these two stories of the night. The younger man dreams of a God who is full of promises and who will help him find rest. The older man wrestles with a God who won't even reveal his identity. The older man has found a deeper faith, one with fewer answers and that leaves him limping.

Our little boy won the name of Jacob because of this. We named him after the guy who slept by the road and was discovered there by something larger than life, something that became harder to name as he got older. Our Jacob's name is a prayer that he will never stop grappling with life's mysteries, that he can be still long enough for the spirit to find him, that he can stop moving long enough to be moved.

I have watched our Jacob searching for sleep in the wee hours. He thrashes about, twists himself, raises an arm and gets himself in a headlock. It goes on and on. He is wrestling. Finally he throws in the towel and his face relaxes into a peaceful surrender, and I tiptoe out of the room, afraid of disturbing his rest.

2:35AM

[2007]

Every night for the last twelve years, Anne has slept by the phone. She suffers from multiple sclerosis, a cruel illness that restricts her mobility and the symptoms of which, for her, include fatigue. But that is not the reason for the phone. Anne waits for calls. A few minutes after half past two in the morning, perhaps, the phone will ring.

Sometimes the caller is a nuisance, which is why Anne prefers to keep her last name private. But more often than not, somebody is calling for a few words of reassurance. They ask Anne a straightforward question: Am I still alive?

Anne tells them that, yes, they are alive, and then they hang up, released from their confusion, at least for tonight.

The callers have narcolepsy, a debilitating malady best known for its more public symptoms. People who live with narcolepsy are prone to sudden attacks of sleep, often at inconvenient times, such as when they are under a shower or at a staff meeting. Narcolepsy doesn't always give warnings of the onset of sleep; there is no dusk before dark. The waking state just goes off like a light.

The other public symptoms of narcolepsy include extraordinary sleepiness during the day and a condition known as cataplexy, the real hallmark of the condition, in which some of a person's muscles suddenly lose their strength and that person may just fall to the ground, lose grip on something they are holding, or perhaps have their facial muscles fall into an appearance over which they have no control. These situations may last minutes or seconds depending on the individual. They can be triggered by emotional surges, such as laughing at a joke or becoming angry.

In addition to this, there is a private side of narcolepsy that is even more frightening.

One of the features of narcolepsy is the ability to slip straight from wakefulness into REM sleep. This can happen in other situations, such as with tired youngsters or those who have had sleep apnea for years and finally discover effective treatment. But it is an aberration. For most people, before midnight there are four distinct stages of sleep that precede REM sleep, and each of these is characterized by a different pattern of brain waves. These four stages culminate in "deep sleep," a period in which the brain is least active and in which sleep takes over from the brain as the boss of your life. In this state, sleep regulates a number of tasks that need to be accomplished after the wear and tear of another day. That busy beaver known as the cerebal cortex has a break while growth hormone and melatonin do you favors that you would only mess up if you had any say over the matter. You don't dream in deep sleep. Your mind gets out of the road so that your body can look after itself. Deep sleep is the healthiest part of your day, the time when you've left the room so that your body can talk about you honestly.

About an hour and a half after falling asleep, a profound change comes over the nature of sleep, marking the arrival of REM sleep. In this state of rapid eye movement, it is the frenetic activity of your eyes behind closed eyelids that indicates the brain is tired of being stuck on the sideline and wants to start playing again; more than that, it wants to captain the team. In deep sleep, your brain is still while your body is quiet and stable with slow regular heart rate and low blood pressure. In REM sleep, however, your body is physiologically active, and your brain gets restless. Most dreams, although not all, take place in REM sleep when the body is paralyzed, as a natural defense mechanism to stop us acting out our dreams, one of nature's really good ideas. REM sleep is so different from the other parts of sleep that even as nonscientists we can think of three states in our human life: wakefulness, sleep, and REM sleep. It may seem strange that REM was not clinically observed until 1953. But there is a simple reason why a species so obsessed with itself seems to have overlooked such an important part of its daily behavior for forty thousand generations: everybody is asleep when it happens.

There is fierce debate over the function of REM sleep, an experience that is by no means unique to humans. Some of these views we have already heard, and there are people who believe REM is the heart of sleep, the culmination of the four stages that came before it, the crucial factor in the consolidation of memory—although the relationship between REM and memory is a particular bone of contention. Francis Crick—one of the discoverers in 1953, along with James Watson and Maurice Wilkins, of the double helix of DNA— proposed in the 1980s that "in REM sleep, there is an automatic correction mechanism which works to reduce . . . possible confusion of memories." In other words, he saw REM sleep as integral to the working of the mind's elaborate filing system, also known as memory. Others believe that REM has more affinity with wakefulness than sleep and is, in fact, a preparation for wakefulness.

Narcoplepsy allows the rest of us to understand what can happen when the cycle of sleep gets disrupted and for some reason REM sleep decides to come before the other stages. People with narcolepsy may well be in the REM stage within minutes of falling asleep.

One of the symptoms of narcolepsy is sleep paralysis, a situation in which a person is awake, conscious, and their mind is fully active. But they can't move. For a few terrifying moments, the body and brain simply don't seem to connect with each other.

The other private symptom is hallucination. The immediate onset of REM can bring immediate dreams. The problem is that without the slow process that usually leads to REM, a person with narcolepsy can have genuine problems telling if their dreams are real or not. They may wake up in the morning and head off to appointments they have only dreamt about or start looking desperately for the keys of cars they have only dreamt they own. To make matters worse, the dreams are often nasty.

That is why they might ring Anne in the early hours of the morning. They have just dreamt that they are dead. They need an outsider to tell them this is not true.

Anne is a nurse who first encountered narcolepsy when she was training in 1956; one day she found a fellow nurse propped up against a wall, able to hear but not move. She later found herself married to a schoolteacher who has narcolepsy, a condition that isn't life threatening but that has treatment rather than a cure. It is a long-haul illness. Anne's husband waited for ages before he found a doctor who was able to respond appropriately. In the meantime, he would come straight home from school and fall asleep. He would then need another nap later in the evening to wake himself up enough in order get himself to bed. Once he was in bed, he also had PLMD (periodic limb movement disorder), which meant that his night's sleep usually cost Anne a few bruises. Some of the couple's children also have narcolepsy; a genetic factor has often been observed, but the condition is not inevitable. It is possible for one identical twin to have narcolepsy and the other not.

Part of dealing with all this has meant, for Anne, being available to help others. "A woman rang me at some ungodly hour, just after half past two," she says. "There was an angel in her room so she needed to know if she was dead yet. I told her she wasn't and she said that was fine, but she sounded slightly disappointed. She said that it had been very pleasant flying around the room."

None of this experience has led Anne, a devout Lutheran, to doubt her faith in her own dreams: "Ever since I was nine or ten years old, I have been getting dreams from God in which I have been told to pass on messages," she tells me. "On five occasions I was told to inform my aunt that she was pregnant. Twice she miscarried but three of those children are still alive. When my mother-in-law died, she hadn't even been sick but the moment the phone rang I knew what it was. Some people are just given this biblical gift, and I am one of them. I don't know how it happens, but God uses me. I come out with my voice, but it's God talking. So when I am counseling, I will be guided in certain ways. I am guided to what God wants me to say."

Anne does not answer the phone between dawn and noon.

Narcolepsy is rare, although the incidence of it varies from country to country. It is more common than leukemia and affects a similar percentage of people as Parkinson's disease, at least one in two thousand. It was initially thought to be far more freakish, and like many sleep disorders, narcolepsy has a long history of not being taken seriously. It was first marked in the scientific record in 1881 by a French doctor, John-Baptiste-Edouard Gélineau.

Born in 1828, Gélineau became a naval doctor and had some risqué adventures in the Indian Ocean, before settling down to private practice in a rural community, where he had time and space to indulge his interest in natural history. He made his money when, in 1871, he developed a heady brew of bromide, antimony, and picrotoxin, which he marketed as a cure for epilepsy and sold in tablet form. The success of the product probably says more about the desperation, at the time, of families living with epilepsy than it does about the efficacy of the pills. The history of sleep medicine is likewise full of wonder cures. People will pay almost anything for a decent night's sleep.

Gélineau went on to establish a neurological clinic in Paris. One day, a 38-year-old approached him with a bewildering problem. The man was a vital member of the local community; he sold wine barrels. But this active businessman was prone to sudden episodes of sleep in any situation; he also had a proclivity to fall down for no apparent reason, a condition Gélineau called astasia but which we know as cataplexy. Gélineau's observations were astute. He noted that astasia was different from epilepsy in that an epileptic seizure tended to make muscles tighten and contract. Astasia had the opposite effect; the muscles turned to jelly, and furthermore, the attacks seemed to switch off as suddenly as they switched on. He rightly observed that these attacks seemed to follow occasions of strong emotion which, for a French wine barrel merchant, were not infrequent. Gélineau was also correct in his deduction that the problem was somehow located in the brain. In 1880, he coined the term *narcolepsy* to describe the whole complex. But the medical mainstream wasn't especially interested in the findings of a maverick, and it wasn't until the discovery of REM in 1953 that his work was given credit. Meanwhile, Gélineau had found there were other ways to help people sleep. In 1900, he returned to Bordeaux to make wine.

Elizabeth Hickey became aware of the symptoms of narcolepsy following a bout of glandular fever when she was sixteen years of age. But her struggles with the condition go back even further. At school, she had trouble concentrating and was often in trouble with teachers, who liked her to sit near the front of the class so that they could keep an eye on her. This old-fashioned approach to discipline may have some valid basis: by creating mild stress for the pupil, a teacher may be helping them release enough adrenalin to stay awake. Elizabeth's behavior might now be described as attention-deficit/hyperactivity disorder (ADHD); she believes her brain was doing gymnastics to compensate for poor sleep and to keep her awake during the day.

"I am just so thankful that nobody in my family treated me like I was a bad person, so my self-esteem never suffered," she says. "Looking back, that was a huge positive."

Elizabeth is middle-aged now. Though she still has narcolepsy, the drug that enables her to function is Ritalin—the same drug that can arouse controversy for its use in treating children with ADHD. There's a theory that ADHD affects poor sleepers who get through the day by having numerous very short "micro-sleeps," which may last only seconds but which are long enough to rupture concentration and reduce the world to fragments. There are others who'd say that poor sleep is a symptom rather than a cause of ADHD.

Elizabeth showed all the classic signs of narcolepsy. At the age of eleven, she surprised herself by being able to swim across a pool. She was delighted by her achievement and the emotion triggered a cataplectic episode as a result of which she lost the ability to move her limbs, and but for the vigilance of a friend who fished her out, she would have drowned. At the age of sixteen, she had a summer holiday job in a doctor's office where she saw a man whose arm had been nearly severed in a car accident. The shock sent her into a paralysis that lasted five or six hours.

Years later, she found herself as the mother of twins living in Toowoomba in regional Queensland and struggling to cope with profound exhaustion. She was so sleepy all the time that she found it hard to function, but she was inclined to attribute this to the demands of having two babies in the house. One day, some friends from Brisbane arrived unexpectedly on her doorstep. The sheer delight of seeing them once again activated her cataplexy and she fell to the floor.

Cataplexy is just one of a suite of symptoms. As a girl, Elizabeth had an extraordinary number of imaginary friends, each of which was absolutely real to her, and now wonders if this was a result of narcoleptic hallucination. Even as an adult, Elizabeth suffered with hallucinations; she can well recall driving with her children in the backseat and experiencing terrifying mirages of oncoming trucks. She has experienced sleep paralysis and, with that, the most frightening dreams.

"I dreamt that things were eating me up. That I was being interfered with," she says. "It was horrible. The dreams would come with a real, physical pain."

On top of that, Elizabeth would just fall asleep. Anywhere at all. At any time. She had to give up the piano because she used to just fall asleep at the keyboard.

Elizabeth's suffering was intensified by the inability of anyone, including herself, to recognize the problem for what it was. When she was thirty-one, one of her sisters-in-law suggested that Elizabeth's inability to stay awake might be due to more than just her little twins. She got a referral to a neurologist, but the very visit to the surgery caused sufficient adrenalin to be released that she was more awake than she had been for ages and the doctor ended up wondering why she had come to see him. Elizabeth says she then went into denial for a further eight years. Overall, it took twenty-five years of groping in the dark, of feeling like she was living underwater, until, at the age of forty-one, another doctor was able to scratch the word *narcolepsy* into her file.

Narcolepsy is, admittedly, not always an easy diagnosis to make and not one to reach in a hurry. One of the tools in diagnosis is the multiple sleep latency test (MSLT), which needs to be performed in a sleep lab during the day. In this test, the patient is wired up to an electroencephalograph (EEG) and asked to fall asleep, a number of times, during a period in which the person would otherwise be up and about. The test can see how readily the patient falls asleep, a phenomenon that is called *sleep latency*, a simple but effective measure of sleepiness and hence an important clue to sleep deprivation. The EEG can also see if REM sleep jumps the queue and pushes its way into the initial stages of sleep, causing no end of trouble. But the MSLT is fallible, partly because it takes place in such odd conditions for sleep. It is possible to do well on the test and still have narcolepsy. It is equally possible to have early-onset REM and not have narcolepsy.

For Elizabeth, family and community have been significant in enabling her to live more comfortably with the condition; for her, community has meant church community as well as the community of other people forced to share their lives with narcolpesy. Both diet and medication have also played a major role. For various reasons, weight gain is a factor in a number of sleep disorders.

Current thinking is inclined to attribute narcolepsy to low levels of hormones called orexins (also called hypocretins), which are produced by a part of the brain called the hypothalamus, an amazing little gadget that is a bit like the brain's brain. It regulates sleep, appetite, and body temperature, three aspects of our lives that are closely related. Disturb one and the chances are you will disturb the other. Orexin levels are often measured by a spinal tap, an intrusive procedure that can have slight risks of its own. Certain drugs that can help with narcolepsy, such as modafinil, have been found to stir the orexins. But Elizabeth uses large amounts of Ritalin and has long argued that the maximum recommended doses of this drug are nowhere near sufficient for her needs. "If I don't take enough," she says, "I may as well not take any."

Ritalin is not a substance to be toyed with, and Elizabeth is well aware that there is a list of drugs, including some antidepressants, with which it has a dysfunctional relationship. Nevertheless, although Ritalin is often used to keep people awake, for her it means she can get into the depths of sleep without having to wade through shoals of hallucination. At the end of the day, she finds it is important she gets to bed before the effects of the drug have worn off. Otherwise, she can end up spending several hours in the kitchen, pottering around with no idea what she is doing or why. This is called *automatic behavior*, a form of waking sleepfulness that can beset anybody but is part of a range of sleeping disorders, especially narcolepsy. At this stage, only further medication will get Elizabeth to bed. She needs to wake up enough in order to sleep.

It's been a long journey for Elizabeth and thousands like her. She wanted to be a doctor and did well enough at school to qualify to go to medical school. But in the final years of high school, she was already falling asleep so constantly that she doubted she could cope. She took a degree in pure mathematics and psychology instead.

Still, self-doubt has plagued her nearly as intensely as her nightmares. "My journey has been one of self-acceptance," she says. "Before I was properly diagnosed, narcolepsy just took me over and engulfed me. There was nothing of me left. Then I met people with the same disorder who were really learning to live their lives. Now I can see myself as a person who just happens to have a disorder. I am a person. Narcolepsy is a disorder. There's a difference."

3:05 AM

[2005]

The middle of the night is a desert in which mirages appear. One of the mirages is TV.

When Jacob and Clare were little, we used to wonder how such small people could make so much noise. We tried a process called *controlled crying* (often known as Ferberizing), which basically means that you don't pick up the child as soon as they start to cry but wait for a while—even a long while—in the hope that they will learn to comfort and settle themselves.

Controlled crying is not the right name for such uncontrolled misery. We started out with two babies crying. Before long, this had increased to two babies, a toddler, and two adults crying. The only positive is that the sight of mom and dad sitting on the living room floor and sobbing fit to break the drought did, at one point, create a thirty-second hiatus. Perhaps the little ones were simply stunned.

Sometimes we found ourselves perched in front of the screen in the small hours of the night looking at a TV wasteland that made the drought-stricken properties of some of our neighbors seem lush. Jenny would strap herself into a device called a twin feeding pillow, and Jacob and Clare would then be fitted around her like lifeboats on the side of a ship. I would sulk in the background looking like a martyr but not helping very much, except to offer a cup of tea, which I knew Jenny would refuse because the twin feeding pillow could become pretty choppy and we didn't want to risk hot tea spilling onto the lifeboats.

It was during this time that I became acquainted with the spectacle known as the late-night infomercial.

Advertisers know all about the quick hit, the sudden jolt that can crumple the soul on impact. Apparently, all you have to do is look at a billboard for three seconds or an image in a magazine for five seconds and then you're hooked. But there is a special skill in the long, slow campaigns of attrition, the ones that grind you down over a grueling thirty minutes on TV, a period so long that you need to have advertisements within the advertisement to break it up. The ads within the ad are for the same product but are made to look and sound a bit different. You think it's a respite, but the toll-free number at the end is just the same as the one that has been burning into your skull for the last quarter of an hour.

In our late-night sessions, Jenny and I saw an infomercial for an acne treatment that was so desperate for material it included an interview with a pimple, which vowed and swore that nothing brought fear to its heart more than the name of the stuff we were being asked to buy. We also saw an advertisement for a pillow that seemed to go out of its way to be as boring as possible because its basic inducement went something like, "Wouldn't you rather be tucked up in bed being entertained by your own dreams than listening to this drivel?" The pillow, we were told, was allergy-free and guaranteed to help in recovery from a long list of medical conditions, which, in a way, was a legitimate line of argument because it's hard to think of any medical condition that is not helped by a good night's sleep.

The pillows were sold as a pair. They had speakers inside, so you and your beloved could listen to music of which some sultry examples were supplied. Or, it was suggested, you could deal once and for all with the snoring of your partner by listening to recordings of the sound of the sea or of birds or of rain pattering gently on a tin roof, all of which were available on CD from the same toll-free number. I wondered aloud if a CD of falling rain might be a nice Christmas present for our drought-affected neighbors.

One night after Clare was sick, she was up for a middle-of-the-night feeding. She hadn't been eating much, so we were glad that now—at five past three in the morning—something was beginning to go into her belly. Our DVD player, which sat under the TV, displayed the time in green, so we could watch the minutes tick past as, for half an hour, somebody tried to sell us a set of knives.

These knives could do anything, it seemed. In half an hour, you don't just get to see them slip through ripe tomatoes and frozen steak. We saw them make light work of leather, small branches, and a phone directory (although it was never explained why anyone would ever need a kitchen knife to deal with these aspects of reality). Then the big promise: in a minute, we would get to see the knife cut through a tin can.

"Couldn't a can opener do the job?" I asked aloud.

Clare was still feeding.

Before we could see the trick with the can, we were told to get a pencil to write down the toll-free number. We were also advised to have our credit cards ready. The TV had assumed we were now under a spell. We would do whatever it asked.

We saw the knife chew through a tin of soup whose contents spilled onto the demonstration bench.

"That wouldn't happen with a can opener," I said. "With a can opener, you need to keep the can upright."

Jenny ignored me, her attention on Clare.

Next, the knives went out on the streets, which seemed to me a risky thing to do. The proud demonstrator approached strangers with a can of beans and a knife and let them try for themselves. (I can't help but worry that people have died making these advertisements.) Male and female, young and old, the passersby were all duly impressed. One of them said that his grandfather used to open tins of beans with a rifle, so perhaps the knives were a step toward a more peaceful world. Never did anybody mention the fact that a can opener might achieve the same end.

A bit of color was returning to Clare's cheeks.

"It won't be long before she's ready for baked beans," I said.

Jenny looked at me with concern, worried that I was about to pick up the phone.

"Don't worry," I said. "I won't be buying knives. We already have a can opener."

Suddenly, the TV played its trump card. It said that if you rang now, there were people waiting to talk to you. It didn't say they would take your money and send you knives that you may not need. It didn't say that the number to call just happened to be the same number you call if you wanted acne stuff or pillows or, God knows, pogo sticks or treadmills or CDs of the greatest-ever country hits, although they never say what country. The promise it made was that, if you rang now, you wouldn't be speaking to a machine. You would be chatting with a real person. By now it was obvious that there are people so desperate for company in the middle of the night that they will buy pillows, knives, and acne stuff just to hear a friendly voice.

Insomnia is a lonely place. In the middle of the night, people can be as vulnerable as a sick baby. And there are others who are happy to pick over their bones.

"Clare's almost finished," said Jenny. "She's falling asleep. Look."

I was distracted now by all the free stuff that came with the knives. I was starting to change my mind. Maybe this wasn't such a bad deal after all. If you bought the knives, you got a whole battery of kitchen gadgets as well.

"Hey," I said. "You get a wall-mounted can opener as well."

At that moment, Clare vomited all over her mother. Half an hour's work was now all over the sofa.

3:15AM

[2014]

There is a whole lexicon of words for nighttime fears.

Here are a few:

Clinophobia: fear of going to bed

Eosophobia: fear of dawn

Hypnophobia: fear of sleep

Insomniphobia: fear of not sleeping

Noctiphobia: fear of the night

Nyctohylophobia: fear of forests at night

Oneirophobia: fear of dreams

Oneirogmophobia: fear of wet dreams

Optophobia: fear of opening your eyes

Siderophobia: fear of stars

Somniphobia: fear of sleep

There are plenty of us who hold our fears tight, cuddle up with them, take them to bed with us. In some ways, the disrupted sleep of the world is a reflection of the levels of anxiety people take to bed with them.

One of the less commonly acknowledged resources for improving sleep (and relieving anxiety) is the conscious practice of surrender or letting go. There used to be an old word for this. It was *forgiveness.* It is another lost art. It can be a difficult medicine that leads to improved health.

There is a wonderful sequence toward the end of Homer's *Iliad*, an epic about the Trojan war. *The Iliad* is so distinct from *The Odyssey* that scholars debate whether or not they could come from the same source. These works were formed over centuries, although one devotee of

Homer, Adam Nicolson, has pointed out that however much they vary in style they are held together by a guiding vision:

> Homer reeks of long use. His wisdom, his presiding, god-like presence over the tales he tells, is the product of deep retrospect, not immediate reportage. His poetry . . . is also driven by the demands of grief, a clamouring and desperate anxiety about the nature of existence and the pains of mortality.

Homer, whoever or whatever he may have been, created works that were centuries old before even Socrates started teasing his fellow citizens of Athens. But they have an understanding of humanity that keeps them well and truly alive.

Near the finale of *The Iliad,* the Greeks have been camped out for ten years, and the Trojans have spent the same time stuck behind walls. Eventually, Achilles, the Greek hero, does battle with Hector, the son of Troy's King Priam. Achilles kills Hector and decides to drag his body around and around the walled city to taunt the Trojans. He does this for days on end. Priam does not sleep during all this time. He stands on the battlements of Troy, keeping watch over the degradation of the body of his beloved son. He is exhausted in every way.

Then a curious thing happens. Priam steps down from the regal trappings of his position to assume an unfamiliar role, that of a simple father. He leaves a role and becomes a human being, making his way to the tent of mighty Achilles to ask for the body of his son. He expects to be killed. But Priam has not reckoned on the fact that the whole world is now tired of conflict and aggression. It is heartily sick of the macho posturing and muscle flexing of men. It needs something different, and humility is suddenly powerful.

"Wow," interrupted Mpho Tutu from the other side of the world. "That is a wonderful story," she continued after a pause. "I am so full of admiration for that young man."

Later she commented that the road to forgiveness is all about finding strength and finding your voice, exactly as Benedict did. I went home at 3:15 AM, full of gratitude. I checked on the children before I turned in. Ever since he had made that video and shown it, Benedict was happy to sleep in his own bed. His thrashing in the night had come to an end. There was peace at the end of an honest road.

At the key moment of *The Iliad*, one of the pivotal points in all of literature, Achilles offers Priam a souvlaki: lamb, roasted on a rotisserie spit, cut into pieces, and served in bread. The souvlaki is hardly a royal feast. But it does what the sharing of food and hospitality has done for ages. It creates healing and community. The souvlaki is a symbol of forgiveness and reconciliation. Priam finds that his sense of taste, which grief has dulled, returns. Immediately, both Achilles and Priam are ready to sleep:

> Priam broke the silence first:
> Put me to bed quickly, Achilles, Prince.
> Time to rest, to enjoy the sweet relief of sleep.
> Not once have my eyes closed shut beneath my lids
> From the day my son went down beneath your hands . . .
> Day and night I groan, brooding over the countless griefs,
> Groveling in the dung that fills my walled-in court.

Achilles sets up exquisite bedding for them both in the porch, that liminal place between the inner and outer world.

This is not just an experience that belongs in ancient texts. Even closer to home, there was a time when Benny was ten years old when he was plagued by the most devastating anxiety. He was sleeping badly, often turning up in our bedroom at ungodly times. Indeed, there was a period in which we had to get a secondhand mattress and put it on the floor of our room, because Benny simply couldn't sleep on his own. A contributing factor was the way he had been bullied at his school. We ended up having to move him to a different school.

Around this time, I took part in a radio program for the BBC (British Broadcasting Corporation) on the theme of forgiveness. The other guest was Mpho Tutu, the daughter of Desmond Tutu, who has long been one of my personal heroes. Together with her father, Mpho had written a work called *The Book of Forgiving*, a result of the journeys both father and daughter had traveled. Among his many commitments, Desmond Tutu had been the chair of South Africa's Truth and Reconciliation Commission between 1995 and 2000, a post from which he saw quite intimately how tangled and complicated forgiveness can be. Yet Tutu writes about forgiveness with no podium, only with deep humility. He talks about the difference between forgiveness and weakness. In fact, they are opposites. He also explains that forgiveness is not a simple story with a beginning, middle, and end. It is often rather more like putting a puzzle together. Forgiveness is one of the most difficult things anybody can do for their own well-being. Both father and daughter have written about the positive impact it makes on insomnia.

Because of the time zones involved, I found myself in the wee hours of a wet and cold night huddled in a radio studio at a quarter past three in the morning with Mpho Tutu in South Africa and the presenter of the program at the BBC studios in London. I was put in a small booth that felt like a confessional. The technician on the other side of the window was eating corn chips, an activity that, because of his work, he had learned to do in complete silence. It was an amazing talent, one whose secret he should share with the universe. It might not lead to world peace, but it might lead to peace on family car trips, a step in the right direction.

"You drew the short straw to get this time slot," he said.

Nothing could have been further from the truth. I would have turned up at any time to talk to Mpho.

It had been a long day. Benedict had struggled to get out of the car and face school. He'd been quiet on the drive there, but once we reached the gates where he was supposed to be dropped off, he fell apart. This was a regular occurrence at the time. He ran through all sorts of emotions, from tears to anger, and tried all sorts of threats, from promising to run away

from home forever to promising to never leave home so
look after him forever. I sat in the car beside him, wonder
happen and how late I was going to be for work or if I was
it at all today. But I waited patiently with Benny, feeling like
on the edge of a precipice. Eventually, after half an hour, his n
and the storm had passed for today. He got out of the car and
embraced by the gates of school. It reminded me of that mome
airport when you see someone go through the gates at customs and
nothing you can do but let them go to find their own way.

That night, I had expected to be talking with Mpho Tutu about what
giveness might mean in a political context as she looked back on twe
years since South Africa gained a new constitution and Nelson Mande
became president. Instead, she got on the topic of family. That was wher
she had learned about forgiveness.

When it was my turn, I told a story about Benedict. He left his first school because of the merciless manner in which he had been bullied. It is painful even now to think about what he endured, and it would be better not to share the details because I'd rather celebrate my son than anything else. Benedict was crushed but, I believe, found a bedrock within himself of great strength. He made an extraordinary video in which he confronted his aggressors with the catalogue of what they had done. In a calm and detailed manner, he told his story and demanded that people listen, something that he had not been prepared to do earlier. The former school wanted a watered down version, but Benedict held his ground. He then had the courage to insist that this was played to the children concerned. He showed great generosity in giving those kids an opportunity to realize their need for forgiveness. He was creative in the way he relinquished some of the burden he himself was carrying.

3:30 AM

[2007]

There are many forms of torture, but two have appetites all of their own. One is to not be able to stay awake. People with narcolepsy and even those with sleep apnea know how hungry sleep can be; it can eat you whole.

The other form of torture is to not be able to stay asleep. Insomnia is a fussy eater. It will calmly let a person stew; it can sit on the end of someone's bed and watch them cook. Insomnia doesn't believe in comforting pillow talk. It can whisper horrible things at half past three. Experts say insomnia is a symptom rather than a disease, but it can be hard to know what it is a symptom of. Sometimes it is a symptom of itself: it is often the case that the worse people sleep, the worse they sleep.

History is a catalogue of the strange things people have done to get a decent night's sleep. Benjamin Franklin was a one-man renaissance and an individual of spare common sense; he embodied what became the governing myth of American independence: that a single person could be an entire culture all on their own. Like Thomas Edison, he was fascinated by electricity. Unlike Edison, he had time for sleep. What he didn't have time for was sleeplessness. His remedy for insomnia was simple: get two beds. If you couldn't sleep in one, then surely you'd be able to sleep in the other. His reasoning was typically transparent: people slept badly because they were too hot. The reason they got too hot was either because they had too many bedclothes or they had eaten too much or both. In an essay entitled "On Procuring Pleasant Dreams," published in 1786, he writes, "Nothing is more common in the newspapers than instances of people who, after eating a hearty supper, are found dead abed in the morning."

In this essay, Franklin recommends having a cool bed handy to hop into. If you can't afford two beds, then you should get up and walk around without your clothes until "your skin has had time to discharge its load" and you can't take the cold anymore. If you do that, Franklin says, "You will soon fall asleep and your sleep will be sweet and pleasant. All the scenes presented to your fancy will be, too, of the pleasing kind. I am often as agreeably entertained by them as by the scenery of an opera."

Franklin says that he learned these lessons from the story of Methuselah, the grandfather of Noah, who holds the distinction of having lived longer than anyone else in the Bible, a work full of folk who were in no hurry to meet their maker. According to Franklin, Methuselah's secret was that he spent most of his nine hundred and something years sleeping out of doors in the fresh air and, indeed, only agreed to his second five hundred years on condition he could continue sleeping under the stars. It would be interesting to know Franklin's source for this information, because the Book of Genesis deals with Methuselah's millennium in a couple of lines as it rushes headlong toward the flood. Franklin condemns aerophobia, a word he used to describe fear of sleeping with the window open. These days the same word can mean fear of flying, as well as fear of airborne germs. Not only do we have more phobia words but the ones we have are working harder.

If only it were as easy as Franklin thought. Those who suffer from insomnia know that the condition can be callous, often defying explanation and thumbing its nose at attempts to deal with it in a reasonable way. One result has been that, for centuries, insomniacs have been an exploited group, and their difficulties have been a goldmine for everybody from hypnotists to pillow makers. Insomnia is worth lots and lots of money; its hostages can be prepared to pay a king's ransom to escape. As long as people continue to be lured by images of perfect sleep, as opposed to adequate sleep, this will remain the case. For centuries, the insomnia industry has been good at selling a product that it can't deliver and that people often don't need anyway. In this, it shares something with the cosmetic surgery trade: there's no such thing as a perfect body and, even if there was, nobody needs one. But it's harder to make big bucks out of reality. Fantasies are cheap to build and easy to rent.

Drug companies are among those who have done very nicely out of insomnia. The trouble is that not everyone has done nicely out of them.

On the evening of September 13, 2007—the night she died—Mairéad Costigan was staying with her parents, Michael and Margaret, in the harborside suburb of Lavender Bay in Sydney, Australia.

"Her name rhymes with lemonade," explains Michael nine years later. "So that became her nickname at school. I think possibly we were unkind to give her a name her little friends couldn't say or spell."

Mairéad, then aged thirty, retreated here occasionally from her place in Paddington on the other side of Sydney's fabled harbor. She had had a busy day; she had applied for more teaching work at the university, she'd bought a top to wear, paid for new glasses, which she would collect later, and arranged some meetings for the following week. It had been a full day in a full life. She was planning to meet her sister for brunch on the weekend, a friend for coffee the following week, and was talking about a trip to London in the near future; she wasn't saying good-bye to anyone. Mairéad had just completed her doctorate in philosophy two weeks earlier, having written a thesis on aspects of justice and politics in Plato's *Republic*. She was a gifted thinker, and her work in philosophy had turned heads. A semester's teaching at the university had proved demanding, partly because Mairéad's main interest was research, but there was only a couple of weeks of that to go; the pressure was lifting. In many respects, it was a life to envy. There was no conceivable reason to let it go.

That night, she watched TV with her mother in her parents' room and then, at 9:25 PM, said she was going to correct some essays. She changed into her pajamas in readiness for bed and wrote an e-mail arranging a work meeting for lunchtime the following Monday, a message that she never sent.

Mairéad had lived with insomnia for a long time. According to her sister, Siobhán Costigan, she didn't have trouble falling asleep, but staying asleep could be a real bother. Nocturnal noise was especially difficult; she had a ritual for closing windows tight before going to bed to ensure quiet. She even made sure there were weights on papers in the house so that they didn't rustle and disturb her.

Mairéad had tried different remedies over the years and, for about nine months, had been taking a drug called zolpidem, prescribed for her at a walk-in group medical practice where she had seen three or four different doctors, all with access to the same records. The practice had kept Mairéad on zolpidem far longer than recommended. Despite advice from friends, she kept going back and getting more prescriptions. She was desperate.

Zolpidem is marketed in Australia as Stilnox but is known in Britain as Stilnoct and in the United States as Ambien. Since it came on the market in the early nineties, it has been a bonanza for the company behind it, Sanofi-Aventis. Over twenty-five years, Ambien became the market leader in the United States, where although it is only available by prescription, it advertises for business in the open marketplace so that patients know what to tell the doctor to write on those little pads at the end of their nine minutes. By 2015, its position was being challenged by Belsomra, but at its peak, Ambien was worth $2 billion a year in the United States alone. It was a very big slice of a very big pie.

Siobhán was concerned when she heard that her sister was on zolpidem. There had been a well-publicized incident the year before when Jon Mark, a 37-year-old male, had climbed over the balcony of his twelfth-floor apartment while sleepwalking and had fallen to his death. Jon, who had been taking zolpidem, had only been married for a couple of months; his wife had been at preschool with Mairéad and her sisters, Siobhán and Sascha, so they were troubled by the story. They were troubled, too, by Mairéad's consumption of the drug. In the time Mairéad had been taking zolpidem, Siobhán and some friends had noticed changes in her personality and behavior: she had been easily confused, jumpy, prone to lose things such as her wallet, keys, or phone, and her short-term memory seemed poor. And, perhaps even worse, her insomnia seemed to be getting worse, not better.

Six days before she died, Mairéad switched medication. She moved to a drug called zopiclone, sold as Imovane, which she took scrupulously according to instructions. When she died, there were precisely six missing from the pack, suggesting to the police that she had done what she was told and taken one a day.

Zolpidem, zopiclone, and the more recent zaleplon (sometimes sold as Sonata) are known, because of their names, as Z-class drugs. Z-class drugs have been seen as successors to benzodiazepines (such as Mogadon and Valium), which appeared in the sixties and dominated the market from the seventies. These, in turn, took over from the barbiturates that were developed in Germany before World War I and became common in forties and fifties; Adolf von Baeyer, whose name survives in that of another famous drug company, invented barbituric acid in the 1860s when, for reasons best known to himself, he wondered what you got when you mixed animals' urine with apple juice. It was over forty years before someone else noticed that this brew made dogs fall asleep. Both "benzos" and "barbies," simple sedatives by the standards of contemporary pharmacology, had notorious side effects, not least their addictive properties. They didn't just help lives, they took them over.

Before benzos and barbies, there was opium, which dealt wonderfully with insomnia, but in addition to being highly addictive, it had the shortcoming of replacing insomnia with chronic sleeplessness. For some people, this wasn't so bad because the little sleep that was produced by opium, often taken as laudanum, was visited by such appalling dreams that users were just as happy to be awake after all. Samuel Taylor Coleridge, the famously garrulous poet, was crippled by opium, although his fellow addict, Thomas De Quincey, author of 1856's *The Confessions of an English Opium Eater*, loathed Coleridge's "eternal stream of talk which never once intermitted" and lamented that Coleridge never shut up, either awake or asleep, so it was hard to tell what state he was in at any time. Shakespeare knew both about opium ("the poppy") and mandrake ("mandragora"), as well as other "drowsy syrups of the world," none of which could help poor Desdemona. Before Shakespeare, there was valerian, used by the Romans and Greeks. Before that, there was bound to have been something else. In short, there has been a long and time-honored quest for sleep aids—one that is by no means over yet.

At the other end of the spectrum, the US military has reportedly been trying to develop medication to enable soldiers to survive for longer periods without sleep, making them immune to the effects of sleep deprivation. Researchers have been trying to find drugs that will enable the human brain to mimic what happens in the brains of birds that stay awake for long periods during intercontinental migration. They have also been experimenting with modafinil, often sold as Provigil, a drug used in treating narcolepsy, hoping that it might offer a key to enabling soldiers to be alert to some degree 24/7. It's an interesting new scientific field, of course, but one that must be approached with a great deal of caution. Bear in mind that the Three Mile Island nuclear disaster of 1979, the Challenger space shuttle disaster of 1986, the Chernobyl nuclear disaster of 1986, and the Exxon Valdez oil tanker disaster of 1989 have all been attributed to various levels of sleep deprivation in key personnel. Sleepless soldiers sitting in front of panels of flashing buttons are a quite scary idea.

The Z-class drugs of the last twenty years sound like a new line from Mercedes-Benz. But they are not without problems of their own. Z-class drugs act fast, which is one of their attractions. And that was certainly the case with Mairéad Costigan on that fateful night.

Soon after getting into her pajamas, Mairéad Costigan got up suddenly and left her parents' apartment. On her way out, she walked past her father who was asleep in front of the TV, something her sister says she would never have done if she'd been conscious because she'd been concerned about her dad's health and would have stopped to check on him and get him to bed.

It was a cool evening in Sydney, the temperature falling to fifty degrees Fahrenheit overnight, but Mairéad was barefoot and only wearing her pajamas. She was normally particular about her appearance, but now her hair was unkempt. The reason was that she was already fast asleep.

She never woke again.

Mairéad walked in her sleep up onto the cycleway of the Sydney Harbour Bridge. Cyclists are possessive of this little piece of turf; pedestrians have a dedicated walkway on the other side of the bridge and cyclists will always bark at stray walkers to get off their side and back where they belong. The CCTV footage of that night shows Mairéad zigzagging along the cycleway; a few bikes slipped past her, but she was oblivious to them and any curt advice they may have sent her way. Soon afterward, Mairéad climbed the chest-high parapet of the bridge, and it was from here that she slipped.

Her landing sixty-five feet below was heard by a homeless man in a bus shelter. A passing nurse tried to revive her, but her injuries were substantial. Had Mairéad been awake, she would have instinctively put her hands out to cushion her landing. But because she was asleep, she didn't make any attempt to protect herself. The impact showed.

It was not yet ten o'clock. When the police called on Mairéad's parents several hours later, her mother, Margaret, responded to the news with one word: "Stilnox."

The toxicology showed no alcohol in Mairéad's system nor any other drug apart from Imovane. She had no history of depressive or mental illness.

Before long, Siobhán Costigan and her sister, Sascha, had flown to Brisbane to speak with Dr. Geraldine Moses, a clinical pharmacist who was the founder of the Adverse Medicine Events Line, a phone-in service based at Brisbane's Mater Hospital and funded by the National Prescribing Service. The line began as a port of call for people who were experiencing difficulties with medication. It gives advice but also listens to stories; it has become a significant conduit for the grass-roots experience of consumers to reach the authorities such as the Therapeutic Goods Administration (TGA), who regulate the availability of medicines. A third of the calls made to the Adverse Medicine Events Line report unexpected side effects, many of which have not been described in any official literature.

Dr. Moses grew up with a firsthand understanding of the mystique that can surround medicines, especially so-called wonder drugs. Her father ran a pharmacy in the center of Brisbane, where he became renowned for his hangover cures. After a big night, headsore people, mainly blokes, would make their way into the city from the suburbs in search of Mr. Moses's ministrations. There was something about the father confessor in Moses's approach; he kept a special section of the pharmacy where he could sit people down and talk to them.

"Dad's magic hangover cure was really just codeine," explains Moses. "But the tablets were red. The customers had confidence in them because of that. Dad said that people believed that all good medicines were red."

In her own career, Dr. Moses found she had a gift for explaining how medication worked in a clear and accessible manner, cutting through the aura that can hover around those blister packs. In the mid-nineties, she had a popular segment on national talk radio doing just that. It is still a widespread need; Australians take a lot of tablets.

When Dr. Moses started hearing stories about the Z-class drugs such as zolpidem, she initially dismissed them. The reports were too outlandish. Besides, there were plenty of people who loved these drugs and were grateful to them. But the stories just kept coming, each one as improbable as the one before it. Most of them concerned what are known as parasomnias, the name given to things people do when they are asleep that they should only do when awake. Sleepwalking, sleeptalking, sleepdriving, sleepironing, sleepcleaning, sleepsex, sleepcooking, and sleepeating are all parasomnias. So is sleep carwashing. So, unfortunately, are various forms of sleep violence, including violence to oneself.

Here's a sample. A man got in his car and drove hundreds and hundreds of miles to the house of friends, where he had a cup of tea in the middle of the night. When the friends rang the next day to check he got home safely, he had no recollection whatever of having made the trip. Another man, who slept in the nude, found himself in his car at a service station about to fill up with gas. He then realized he had forgotten his wallet. He had forgotten his wallet because he had forgotten his clothes.

These stories aren't really funny. A woman needed half her leg amputated after she slipped and broke the leg as she was cleaning her bathtub while she was still asleep. Not even the pain of a broken bone woke her and the angle at which she fell cut off circulation to the leg, killing the limb. When she finally awoke, she was close to a multi-organ collapse. Another woman mysteriously gained almost 95 pounds; her partner confronted her with evidence of her nocturnal cooking extravaganzas and she was dumbfounded. In April 2008, a judge in Sydney accepted that Robert Kingston, who was involved in a traffic accident when he was driving on the wrong side of the road in his sleep attire with a blood alcohol level of 0.105 percent, may have been acting strangely because of the Stilnox in his system. By that point, Australian authorities had received over a thousand reports of bizarre reactions to Stilnox, 10 percent of them related to driving.

Worst of all is an increasing list of people who have managed to kill themselves while asleep. Dr. Moses says that her modest help line has already received many such reports. Once the story of Mairéad Costigan received publicity, another family from Western Australia made contact to talk about a similar tragedy that had befallen a young woman there. These stories do not include the near misses: the man who texted a good-bye message to his family from a beach in the middle of the night and awoke in the hospital; the woman who pointed a gun to her head and was just lucky the gun jammed. The one and only constant in all these scenarios is the presence of a Z-class drug.

Geraldine Moses attributes the problems to the sophisticated manner in which the drugs work. The older generation of sleeping tablets are simply sedatives and help at bedtime by creating calm. But zolpidem actually changes the architecture of sleep. In Moses's words, it disturbs the great symphony of sleep.

Zolpidem works by stimulating the pathways used in the brain by a hormone called dopamine, produced in the hypothalamus. Dopamine has a job description as long as your arm but helping to organize sleep is part of it. In improving traffic flow for dopamine, zolpidem extends the time a sleeper spends in stage-three and stage-four sleep and delays the onset of REM sleep. It thus alters the natural progression of the sleep cycle and blurs the boundaries between REM and non-REM sleep. For some people, this means an increased likelihood of parasomnias, such as sleepwalking, taking place during a stage of sleep when the body is not paralyzed in the way nature usually arranges to prevent us acting out our dreams. For a significant number of those people, the results have been tragic.

As well as negatives, zolpidem has had some unexpected positives as well. These include a case reported in South Africa of a young man, Louis Engelbrecht, who spent three years in a coma after the bicycle he was riding was hit by a car near his home; he seemed unable to respond to any communication from outside himself. His mother, Seinie, noticed that he had become increasingly restless, tearing at his own bedding, and so his doctor, Wally Nel, prescribed zolpidem for him. A coma is not sleep (a key difference is that sleep is easily reversible), so the idea of giving sleep medication to someone in a coma is not as bizarre as it sounds. Within half an hour, Louis had spoken to his mother for the first time in years. The drug is now being used on a range of people suffering brain injury, with a reasonable level of success. Zolpidem has also been found helpful to patients with Bell's palsy, Parkinson's disease, and even restless leg syndrome, all of which is good news. But you do have to scratch your head. All drugs have side effects and are inevitably released onto the market, and become part of millions of lives without the manufacturer knowing everything they do. Indeed, says Geraldine Moses, all kinds of tests are done on new drugs before they are released, but their impact on sleep is seldom investigated.

Six months after her sister's death, Siobhán Costigan had given up most of her work as a graphic designer to spend time getting to the bottom of what happened to her sister and to publicize Mairéad's experience with Z-class drugs. The Costigans were among those who campaigned in February 2008 to have the drug reclassified, meaning that it would be categorized alongside drugs that are much harder to access. The drug wasn't reclassified, but there was a warning added to the product that advised of "potentially dangerous complex sleep-related behaviors," ensuring that both patients and prescribers were more aware of its track record. For Siobhán, this was at least a step in the right direction. She points out that the official body in Australia responsible for the reclassification is a government agency funded by the pharmaceutical industry; companies pay fees to have drugs registered.

Sanofi-Aventis has responded to adverse publicity with a poker face. A communications representative said that the company had been inundated with calls from people who had benefited from the drug but that the media wasn't interested in them. In a statement released in April 2007, the company said that problems with the drug had usually resulted from its improper use, especially taking it with alcohol. This was assuredly not the case with Mairéad Costigan. Consumer information for the drug published in June 2007 included a list of side effects at the beginning and end of which, in bold type, was the following advice: "Do not be alarmed by this list of possible side effects. You may not experience any of them." The last point on a secondary list of "less common adverse effects" is "sleepwalking or other behaviors while asleep." Blink and you'd miss it.

In November 2016, Mairead would have turned forty. Her family still grieves for her deeply. Nine years after Mairead's death, her father

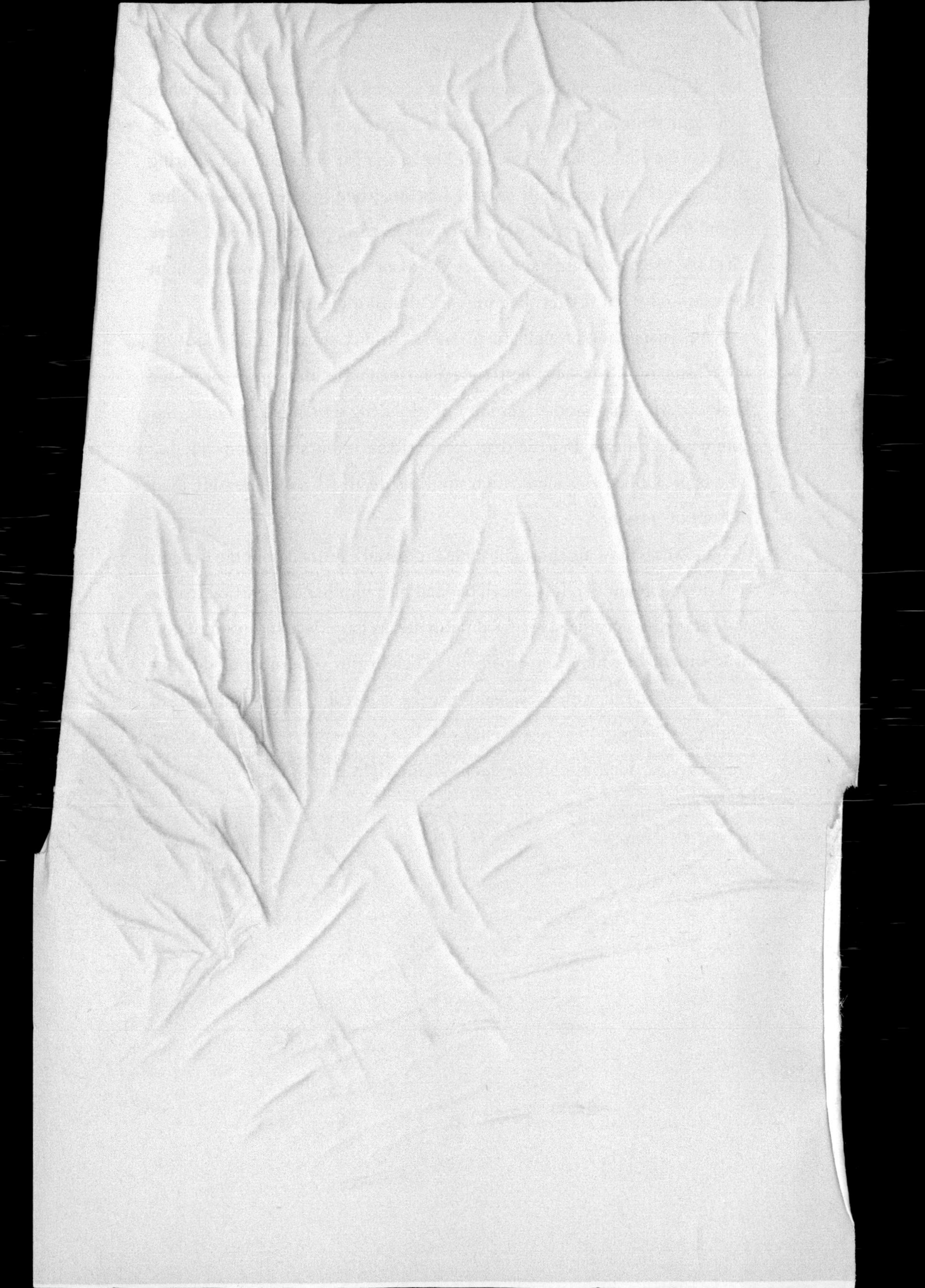

says, "I live with the aching memory of several worrying moments
I thought there may possibly have been warning signs about the
These were times when I was with her in the car and elsewhere du
the last weeks and especially days of her life, when I sometimes drove
to and from the university, where she was lecturing part-time and whe
I had the use of an office as adjunct professor. I have tried to write abo
these episodes but the recollections are almost too painful to bear."

Nine years have brought many more Stilnox stories to the door of the Costigans. They have befriended, for example, the family of Peter Dickson, who was another casualty of the drug. His mother, Jenny, has commented on social media that Peter walked in his sleep from the day he was prescribed this medication until his death six days later. He died in front of a train.

Nine years have also brought some increased caution about the drug in the wider community. It has been banned by a number of sporting teams, and these decisions have received publicity, as has any association of the drug with bizarre behavior, especially in celebrities. A number of doctors are wary of it. But with a few small tweaks, it is still doing great business. Nothing substantial has really changed. For one reason or another, there are plenty of doctors who are sleepwalking right past the warning signs.

3:33AM

[1860]

Everyone sleeps, even those who claim not to. Sometimes people don't know they are asleep because, well, they sleep through it. For many years, the record for sleeplessness was held by a Californian teenager, Randy Gardner, who, in 1965, managed to stay awake for eleven days—a grand total of 264 hours. Gardner achieved this feat with the help of friends and a lot of physical activity including walking, using a baseball machine in a pinball arcade, and playing basketball. He got a bit cranky at different times, lashing out at the people who were keeping him awake because he had forgotten why they were tormenting him like this, and he probably had a lot of micro-sleeps, of which people were not yet aware. But the ordeal didn't seem to leave Gardner with any permanent scars. Once he'd reached the milestone, he slept for nearly fifteen hours. The next night he slept for over ten. After that, he seems to have returned to a normal sleep pattern. Randy was most at risk during the marathon itself.

History is full of famous insomniacs. Some have tried to make a virtue of their affliction. Margaret Thatcher, prime minister of Britain from 1979 to 1990, wrote in her memoirs that she only slept for a few hours a night. She wrote, "There was an intensity about the job of being Prime Minister which made sleep seem a luxury. In any case, over the years I had trained myself to do with about four hours a night." Thatcher presented herself as keeping vigil over the fortunes of the nation, perhaps not an insomniac so much as someone, like Thomas Edison, who had more important things to do than sleep. As a matter of fact, she didn't; the belief is a delusion. Voters may have been less impressed if she said she drank heavily but the effects are basically the same as sleeplessness. According to researcher

Paul Martin, somebody who has been awake for twenty-one hours will have the same reaction time and cognitive impairment as somebody with a blood alcohol level of 0.08 percent. They shouldn't be driving a car, let alone a country.

Charles Dickens was a genuine insomniac, and his anxiety concerning bedtime reflected in a curious obsession. He would only sleep in a bed with the head pointing north and the feet pointing south. People have come up with many such rituals, some easier to explain than others. Elizabeth I, for example, always slept with a sword in her bed, perhaps to defend her virginity or perhaps to ward off the armada if it got that far. The purpose of Dickens's obsession may have been the obsession itself: he was equally fussy about dress and food, although beds had a special hold on him. He always had trouble getting a bed to do what it was supposed to do; at least the direction it pointed was one aspect of its behavior he could control. Dickens's problems with sleep are evident in his capacity, on occasion, to blur the line between dreams and reality. Ideas for plots sometimes came to him in dreams, and he called the month's installment, for whatever serialized novel he was working on at the time, "my month's dream."

Like Shakespeare, Dickens wrote a lot about sleep. The reason was that, like Shakespeare, he wrote a lot about everything. They both had voracious imaginations; the whole world was not enough to feed it, so they needed to create extra worlds. They made these new worlds by observing the one they already had. They both described sleep apnea, for example, long before the condition had medical credentials. Shakespeare saw it in Falstaff. Dickens saw it in Joe, a character in the book that made his name—1836's *The Pickwick Papers.* Joe is introduced as "a fat and red faced boy in a state of somnolency who divided his time into small allotments of sleeping and eating." Joe is obese; the only thing that rouses him is having his leg pinched and his appearance greeted by the refrain, "Damn that boy, he's fallen asleep again." Indeed, in those early days sleep apnea was known as Pickwickian Syndrome. Yet an even better description, one which seems to understand the underlying seriousness of the condition, is found in the character of John Willet, proprietor of the Maypole Inn in 1841's *Barnaby Rudge*. Willet's breathing, when asleep, is likened to the problem of a carpenter trying to get through a knot in a piece of timber. On one occasion, Willet "came to another knot—one of surpassing obduracy—which bade fair to throw him into convulsions, but which he got over at last without waking, by an effort quite superhuman." Mr. Pickwick experiences insomnia ("that disagreeable state of mind in which a sensation of bodily weariness . . . contends against an inability to sleep"), while in *Bleak House*, Volumnia Dedlock is among those legions of people who claim not to sleep when in fact they do. There are descriptions of sleep spasms in the character of Twemlow in *Our Mutual Friend* and of sleep paralysis in *Oliver Twist*: "There is a kind of sleep which steals upon us sometimes, which, while it holds the body prisoner does not free the mind from a sense of things around it." A minor character in *Hard Times* takes to their bed for fourteen years, something of a Victorian pastime.

Dickens included sleep in all its guises in his literary works—most likely because he so often and so continually battled with it himself. In many respects, his spirit could find rest only in his fiction, a world of his own wonderful making; reality, on the other hand, was for him a place of profound restlessness, as it had been ever since his childhood. He said, "My own comfort is in Motion"—with a capital M. He was always on the move, an escape artist, a mirror to the world, a gifted mimic, and a generous-hearted man, but one with little capacity for self-reflection. He painted hundreds of vivid characters, his prose using primary colors to subtle effect, but he never portrayed himself, at least not with conviction. The narrative voice in some of his personal writing is among his least sure characterization. Perhaps he knew that if he ever stopped, reality would catch up with him, not least the traumas of his own childhood and its emotional privations. Something was disturbed in his early years that never settled. All his life he was a magnificent observer; he never missed a thing. His imagination was both his blanket and his bed. His books celebrated human entanglement; in private, Dickens was controlling.

Dickens knew one city, London, like few others have known any city. The source for this intimacy was that Dickens knew London at night, in the hours after it took off its makeup. *Great Expectations*, to take a single example, turns on the moment when the convict Magwitch reenters the life of Pip; that scene rests on its depiction of a wild London night. London was Dickens's bed partner as well as his quarry; he loved her and needed her chaos to pillow his own. The pair of them spent many restless nights together, tossing and turning. He called London his "magic lantern."

Because of all of this, it is hardly surprising that Dickens's remedy for his insomnia was to get out of bed and start walking. Early in his career, in the 1830s, he wrote journalism under the name of Boz. A typical piece recommends getting to know London between three and four in the morning, the time when she gives up her secrets: "But the streets of London, to be beheld in the very height of their glory, should be seen on a dark, dull, murky winter's night when there is just enough damp stealing down to make the pavement greasy."

Just under twenty years later, in 1852—a period during which his marriage to Catherine was unraveling after the pair had had ten children, one of whom had sadly died—Dickens wrote a piece called "Lying Awake" for the magazine he edited, *Household Words*. He says that, since he was a "very small boy," he was familiar with Benjamin Franklin's "paper on the art of procuring pleasant dreams," the one in which he urged insomniacs to have two beds. Dickens did not find Franklin's advice very helpful, writing, "I have performed the whole ceremony and if it were possible for me to be more saucer-eye than I was before, that was the only result that came of it."

Dickens runs through the disjointed chain of thought that filled the hours of his insomnia before reaching the conclusion that he would afterward stick to:

> I found I had been lying awake so long that the very dead began to wake too, and to crowd into my thoughts most sorrowfully. Therefore I resolved to lie awake no more, but to get up and go out for a night walk—which resolution was an acceptable relief to me, as I dare say it may prove to a great many more.

Dickens loved the hours when he could have London all to himself, especially the time between when the church clocks struck three and when they struck four. He wrote an article in 1860 for *All the Year Round*, the magazine he edited after he fell out with the management of *Household Words*, about "the restlessness of a great city and the way in which it tumbles and tosses before it can get to sleep." "Night Walks" is a guided tour through London on a cold night in March after the last of the public houses have sent their patrons home. Dickens walks and walks and walks, weaving through the "interminable tangle of streets," stopping outside the Debtors' Door of Newgate Prison ("which has been Death's Door to so many"), a place that had been the source of special fear and loathing in his life. As a church bell strikes three, he stumbles over a sleeping kid of about twenty, and the two are terrified of each other. As another bell peals four, he enters a dark and empty theater and gropes his way to the stage. After four, he could get coffee and toast at Covent Garden, but he needed to wake the stall holder; Dickens remarks that the fruit and vegetables at Covent Garden get treated better than the people who sell it.

There is a touching moment when Dickens stops outside the walls of Bethlehem, a place that has given the word *bedlam* to the language. Bethlehem was a prison that should have been a hospital; it housed people whose minds had frayed to such an extent that even the frayed streets of London couldn't accommodate them. Dickens wryly suggests that he is not much different from one of the inmates on the other side of the wall whom he once met; this man used to believe Queen Victoria dined with him in her pajamas. Dickens finds fellow feeling with the man in the reflection that no mind will do its owner's bidding while the owner is asleep:

> Are not the sane and the insane equal at night as the sane lie a dreaming? Are not all of us outside this hospital, who dream, more or less in the condition of those inside it, every night of our lives.

Tired and ready for bed, Dickens would go home at sunrise and sleep soundly. He didn't lose sleep over insomnia; he used it to spend time with his love, the city.

The anxiety caused by insomnia can have worse effects than the insomnia itself. After centuries of experimenting with drugs, tonics, aromatherapy, and hot bedtime drinks, some of the best advice to those who can't sleep is often to stop trying to sleep and, at the very least, to stop looking at the clock. Insomnia loves attention; deprived of this, it occasionally sulks and goes away. I once knew a nurse who said that patients in the hospital were regularly plagued by poor sleep. It is curious, given the importance of sleep to both physical and psychological healing, that hospitals are among the hardest places in the world to get any. My friend used to walk the wards in the small hours, noticing the number of eyes looking for something to look at. She'd switch on the televisions that hung over the beds, turning down the volume. Twenty minutes or half an hour later, she would go around the ward again: the patients would now have nodded off, and she'd turn off the screens. There are times when the best form of attack is surrender.

4:30AM

[2007]

By the time she was three and a bit, we had realized that Clare was a strong character and a good talker. She went to bed chatting, could often be heard in the night talking to one of her many cuddly friends, and got up in the morning still halfway through the paragraph she had put on hold when she went to sleep. She threw herself into anything she did and sometimes threw toys, books, and furniture into it as well.

One day, Jenny discovered that a well-known ballet school was giving lunchtime performances in a hall not far from our new home. It seemed too good to miss. Our children all loved dancing, even if for the boys, now aged five and three, dancing usually meant getting dressed up as pirates and terrorizing each other with swords. Clare was disdainful of such crude approaches. When she wanted to dance, she put on her pink slippers, her Snow White dress, and (admittedly) her pirate hat and twirled around on the rug in the lounge, indignant when the boys failed to respect her performance space. We were in a house full of pirates. Clare had decided that she was a princess pirate, a higher order of being than a captain or cabin boy, but a pirate nonetheless.

The ballet school was performing scenes from *Don Quixote*, which meant we had to break the news that while Don Quixote fought with windmills he was not, strictly speaking, a pirate as he had a horse rather than a ship. The boys showed great tolerance and agreed to go anyway, as long as they could dress up as pirates. We were a little hesitant because we had recently taken them to see *The Pirates of Penzance* at the school where I worked and they had dressed for the occasion. They loved the show but it was all we could do to stop them jumping up and joining in; Jake cried at the end because he wanted to put it on again and see it all over like a DVD. That's the problem with live theater, same as with live anything. It ends.

Don Quixote is a famous victim of sleeplessness. At the beginning of his adventures, we learn that the lovable knight with the sad face has spent so much time reading books about chivalry, the pulp fiction of the time, that he sits up from dusk to dawn, and as a result of not sleeping, his brain has either dried up or withered (depending on the translation) and he has gone mad, losing the useful ability to distinguish reality from fantasy. The idea of a brain drying up through lack of sleep is resonant. The word "exhaustion" originates in the Latin word *haustus*, meaning "drink," particularly with the connotation of drinking deeply or drinking right up. The suggestion is not that drink leads to tiredness. Rather, being exhausted means that you have been drunk to the lees, that every drop has been squeezed out of you. We frequently meet Don Quixote keeping vigil through the night for some daft purpose or other. Throughout his struggles, people often try to get him to have a lie down and a good sleep, not least his trusting friend, Sancho, who has things to say on the subject, culled from his inexhaustible fund of proverbs and aphorisms. In chapter XLIII, he says, "When we're asleep, we're all the same, great and small, rich and poor." Later, in chapter LXVIII, he imparts this bit of wisdom regarding sleep:

> All I do know is that so long as I am asleep I am rid of all fears and hopes and toils and glory, and long live the man who invented sleep, the cloak that covers all human thoughts, the food that takes away hunger, the water that chases away thirst, the fire that warms the cold, the cold that cools the heat and, in short, the universal coinage that can buy anything, the scales and weights that make the shepherd the equal of the king and

> the fool the equal of the wise man. There's only one drawback about sleep, so I've heard — it's like death, because there's very little difference between a man who's asleep and one who's dead.

In *Don Quixote*, the devil is always tagged as the one who never sleeps, the saboteur of the human spirit. At the end of his days, the knight of sorrowful countenance returns home and his doctor diagnoses depression and despondency. He gets into bed and, at long last, has a decent sleep. When he wakes up, his mind is free and clear, the shadows have vanished and he realizes that he has been living in an artificial world, that all the silly fantasies he has been reading have shut out his light and that only a better relationship with reality can heal him.

I can't recall the precise moment at which Peter Pan entered our lives. But Peter Pan is like that: he slips in through the window at night when everyone is asleep. Peter Pan is so light that he can—as J. M. Barrie, the wealthy Scottish writer who first dreamed of Peter Pan, puts it—"sleep in the air without falling, by merely lying on his back and floating." He is so naughty that he can creep up behind the stars and blow them out. But he became a great friend because he led us to a place that exists somewhere between here and morning called Neverland, a place where a crocodile has swallowed the alarm clock and thus conveniently dealt with the problem of ever having to wake up. It was in Neverland that we met Captain Hook, our first real pirate.

Peter Pan ran away from his parents on the day he was born because he never wanted to become a man; he always has his baby teeth. Barrie seems to have shared something of this: Barrie was always a boy, even to the extent that he seems to have been physically unable to consummate his marriage in 1894 to a beautiful actress, Mary Ansell. They certainly had no children, and the marriage dissolved without much fizz. It seems that the emotional keystone in Barrie's life was a moment when he approached the bedside of his mother, Margaret, who had taken to her bed for long periods to mourn the death of David, a favorite son who had died at the age of thirteen when he hit his head on the ice in a skating accident. Seeing a figure approach her bed in the gloom, Margaret mistook Barrie for the dead David. "No, it's not him, it's just me," said Barrie. But the cogs in Barrie's soul locked in that place, and he soon began to imitate the mannerisms of his dead brother. He remained a boy for the rest of his life. It seems he could not allow his mother to lose another boy by becoming a man. How ironic that such sad and strange circumstances led, however indirectly, to the joy of Peter Pan.

5:10 AM

[2008]

As part of my work at school, I was able to accompany a group of thirty-five sixteen- and seventeen-year-old boys to World Youth Day in Sydney, in July 2008. World Youth Day, held every so often in a different part of the world, is a huge week-long festival that culminates in a visit from the Pope. In the buildup to the one in Sydney, there were concerns that it could be the smallest World Youth Day ever because maybe only a quarter of a million people might turn up, especially given the impact high fuel prices were having on airfares. I wasn't worried. A quarter of a million sounded like plenty to me. I don't much like crowds, and besides, having been a Catholic priest, I have a complex relationship with the church authorities who would be appearing in all their finery. I find it hard to believe that Jesus died naked on the cross so that the rest of us could have a fancy-dress party. I wasn't looking forward to World Youth Day.

The boys from school rescued me from getting too tangled up inside myself. The Pope seemed like a pleasant old gentleman who was quite capable of enjoying himself. I had started to think more warmly of him from the moment he took the same name as we had given our Benedict; I imagined we must have had more in common than I realized. Every time he spoke in Sydney, I found, despite myself, that I was wholly caught up with the simple depth and elegance of what he said. He named things that were important to me.

Quite apart from that, it soon became obvious that the crowd, which ended up closer to 400,000, was really the best part. While there were dozens of cardinals and bishops way off in the distance, all solemnly dressed up and doing as much as any mardi gras to support the dry cleaners of the world, there were thousands of young people much closer to us. The kids had the time of their lives. Their godly joy was infectious. It budged the cranky heart of their middle-aged teacher.

Of course there were hassles. We traveled by bus, and upon arrival at the school where we were booked to stay, we found that the organizers had thought it would be fun to send us and our bags to two different places. Deprived of my breathing machine, I went downstairs that night and curled up beside the photocopier in the staff room. To my surprise, there were times when I had slept worse in my own bed.

It takes all types to make a religion, and the week before the Pope arrived was a spirited one, culminating in a 6-mile walk across the Sydney Harbour Bridge to Randwick Racecourse, where the Pope would be saying Mass the following morning. The bridge is closed on rare occasions; the time before this was for a walk for reconciliation between indigenous and nonindigenous Australians in 2000. The prospect of getting across it without having to pay the toll was too good to miss. Along the way, I thought of my mother who, as a wide-eyed little girl holding the hand of her mother, had walked across the bridge on the day it was opened in 1932. She had walked across it again for its golden jubilee in 1982. These days, Mum wasn't walking far.

I rang her from the bridge; she was in bed watching on TV, annoyed that the Pope was a smoker but pleased that he liked cats. I also called Jenny to reminisce about a time before we were married when we had also walked across the bridge on a blustery day and we were the only people crazy enough to do it. She had her hands full looking after six young people at that moment, our own and three cousins. Her present moment was too full to share with the past.

Along the way, onlookers joined the festive atmosphere. A boy wore a T-shirt that read I MAY LIVE IN MY OWN LITTLE WORLD BUT AT LEAST I KNOW EVERYBODY HERE. Workers on a building site made a cross out of iron pipes and hammed it up for the crowds, which were more appreciative of irony than I had imagined they would be. Pilgrims are people looking for signs. Somebody stood on the wayside with a placard reading this is not a sign.

That night, 200,000 people slept in the open air at the racecourse. To be honest, not all of them slept, at least not until well after bedtime. But by the time I got out of my sleeping bag at 5:10 AM to beat the line for the toilets—facilities whose condition reminded us that even spiritual gatherings have their physical dimensions— the whole crowd had settled. I began to walk around the outer track of the course.

Everybody had been given a shiny thermal blanket to sleep under. These looked like large sheets of aluminium foil and were probably more useful for cooking than a serious attempt on Everest. But most people were using them on this crisp, clear night, and at 5:10 AM, the full moon reflected light onto the peaks and troughs of all these shiny blankets as if they were a sea. I kept walking. There were 200,000 people here, but I had the place to myself. The stillness of the sleeping crowd was one of the most serene and beautiful things I had ever seen. The moon played with the shapes made by the thermal blankets. It was a prayer without words. I thought of the same moon shining on hundreds of thousands of refugees sleeping in a camp on the Sudan border and wondered about life's lottery. I hoped that this sleeping gathering might help to change the odds.

Before long, Jenny rang. She had had a choppy night. There were six young things in the house, three of them homesick and the other three sick of home. We talked until we were ready to laugh about it. We decided that six people can trouble more sleep than 200,000.

6:00 AM

[1851]

As dawn approached on Sunday, August 18, 1851, Honoré de Balzac was dying in his bed. Balzac was larger than life in every sense. He left behind a remarkable legacy, not least the intricate panorama of *La Comédie Humaine*, a canvas of more than a hundred novels and plays that portray an entire culture: France in the turbulent first half of the 19th century. Balzac's great comedy is predicated on the same understanding that all comedy is—that, as he said, "nothing is insignificant." Balzac was so heavily invested in this project that, as he approached death, struggling with considerable pain, he began calling on some of its characters. Victor Hugo, of *Les Misérables* fame, visited Balzac a matter of hours before he died and later described the scene:

> I was in Balzac's bedroom. A bed stood in the middle of the room. A mahogany bed with supports and straps at either end leading to an apparatus for moving the patient. Monsieur de Balzac lay in the bed, his head propped against a heap of pillows to which red damask cushions borrowed from the bedroom sofa had been added. His face was purple, almost black, and turned to the right: he was unshaven; his hair was grey and cropped short, his eyes open and staring. I saw him in profile, and seen thus he resembled the Emperor.

This is a fine tableau from the theater of death. Hugo declares that Europe is about to lose a great spirit: Balzac *in extremis* even looks like the emperor. The irony of all this lying in state is that Balzac had long believed that bed is the stage on which people look most absurd. He thought that if a couple were serious about preserving their relationship, they should never share a bed. The reason is that, when asleep, most people end up "sticking out their tongues at the passers-by" and resemble the gargoyles of Michelangelo. In his strange and wonderful book *The Physiology of Marriage*, Balzac writes:

> If you knew that one of your rivals had found a way of placing you, in full view of the woman who is dear to you, in a situation in which you must appear sublimely ridiculous: for example, with your face all distorted like that of a mask, or with your eloquent lips dribbling like the copper orifice of some greedy fountain—you would no doubt stab him in the heart. Such a rival is sleep. Is there a man living who knows what he looks like and what he does when he is asleep? We are then living corpses, at the mercy of the unknown power that lays hold of us in spite of ourselves, and manifests itself in the strangest of ways; some men sleep intelligently, others like clowns.

Balzac was no prude. He longed for the days when primitive people had sex in caverns, ravines, and caves, a way of life he missed out on by a mere ten thousand years. He resents the fact that civilization has forced a million people to live "shut up in four square miles" and hence to share conjugal beds, an arrangement that threatened conjugality, writing, "Sleep alone and your love will be sublime; sleep in a twin bedstead and it will be ridiculous."

Considering the alternatives, Balzac believed that separate rooms are desirable because they suit both those couples who are so passionately devoted to each other that distance is no obstacle, as well as those who are indifferent to each other, in which case distance doesn't matter. He describes "the twin bedstead"—that is, single beds under the same canopy—as a "Jesuitical way of sleeping" and remarks that in the early days of marriage "the distance between heaven and hell is not more impassable than the space which divides your two beds." He warns that this state wears off and a bed will resume its primary function, namely for sleep—and, eventually, for death. In due course, the Victorian age would become coy about sex. It would compensate with an elaborate pornography of death. Our exhausted culture has swung the other way: it has a tired and restless prurience about sex and goes to great lengths to sanitize death and keep it out of sight. We think we are superior to the Victorians without realizing how ridiculous we look in our own sleep.

When Balzac died, he had just turned fifty-one. He'd crammed a lot into those years: hundreds of books and hundreds of thousands of cups of coffee. Balzac began his addiction to coffee while at school and drank sixty cups a day, brewed as thick as oil. It was coffee that enabled him to maintain his unsocial work regime: he slept in the early evening, rose at midnight (dressed in the monk's robe that became his hallmark), and then wrote in feverish stretches of twelve or fifteen hours, kept awake by caffeine. To some extent, this lifestyle was necessitated by his complex financial and business arrangements. He was often dodging creditors, staving off ruin, paying fortunes to keep up sunny appearances in the shadow of bankruptcy. Balzac knew how to spend money: his bedrooms, extraordinary productions that reminded some visitors of brothels, didn't help. Yet he spent little time in them, saying that "too much sleep clogs up the mind and makes it sluggish." One of his biographers, Graham Robb, describes coffee as "the corrosive fuel of Balzac's fictional world." He depended on coffee and couldn't live without it. But he didn't love it. He described himself as a "victim" of coffee:

> Coffee is a great power in my life: I have observed its effects on an epic scale. Coffee roasts your insides. Many people claim

> coffee inspires them, but, as everybody knows, coffee only makes boring people more boring. Think about it: although more grocery stores in Paris are staying open until midnight, few writers are actually becoming more spiritual.

In the end, it was coffee that killed him. Balzac had a long list of symptoms, many of which would have taken other mortals a much longer time to acquire, but Dr. Nacquart, his physician of thirty-five years, wrote finally that "an old heart complaint, frequently aggravated by working through the night and by the use or rather the abuse of coffee, to which he had recourse in order to counteract man's natural propensity to sleep, had just taken a new and fatal turn." Coffee, it appears, can be lethal.

Coffee is the most popular and readily accessible form of caffeine, a substance that is used on a daily basis by three-quarters of the people on the planet. It's everywhere. You see people carrying their morning coffee to work in paper cups, holding them out like lamps to light their steps, a personal talisman. Disposable coffee cups have become ubiquitous.

It's a funny mark of modern society that so many cities pride themselves on their coffee culture, as if it makes them unique. What they don't know is that so does everywhere else. If you go to Toronto, the people will tell you that the thing that makes Toronto unique is its love of coffee. In Budapest, they will tell you the same thing. Same in Cairo. Same in Seattle. Or New York. Or London. Most cities have virtually ten thousand places where you can buy a cup of coffee. (And that doesn't count the rivers of cola that flow through supermarkets and fast-food stores, nor the caffeine tablets some people wash down with the cola.)

All this caffeine has one primary purpose: to cheat sleep. It turns night into day. Of course, there are social rituals, and it's great to meet people in a coffee shop and even better to escape from them in one. But it's a curious culture that allows you to relax as long as you spend the time loading up on stimulants. Besides, those people being steered to work by disposable coffee cups, those people holding on to a Coke bottle like it's a handle on something, hardly look like they're relaxing. Caffeine is a potent stimulant, a psychoactive drug domesticated to the point that it is used to punctuate the day.

Caffeine was isolated in 1819 by a Swiss chemist named Gustav von Runge, the man who also discovered quinine. He was put up to the task by the German poet and erstwhile alchemist Johann Wolfgang von Goethe, who was at that time such a megastar that even chemists did his bidding. Goethe is best known for his telling of the Faust legend, a story of how the devil makes deals with the sleepless in the dead of night. Caffeine works by blocking the access to brain receptors of a substance called adenosine, an inhibitory neurotransmitter whose job is usually to tell the brain to slow down and get ready for sleep; caffeine cuts that brake cable. Caffeine doesn't do anything about tiredness; it simply disguises its symptoms for a while.

Coffee accounts for half of the 120,000 tons (yes, tons) of caffeine consumed around the globe every year. There are about 60 milligrams of caffeine in a cup of instant coffee and roughly 120 milligrams in a cup of espresso; in other words, the world's caffeine consumption is two trillion cups of instant or one trillion cups of espresso per annum. That's enough that roughly every single person on the planet, including babies, could have a cup of coffee every single day of the year.

Of course, there are dozens of plants that can provide caffeine; among them are tea, cocoa, and guarana, all with various subspecies. The drug is found in many plants because it is a natural insecticide. It evolved to stop the plants being eaten, and with characteristic perversity, humans are among the very few creatures who regard the stuff as anything other than a poison or, at least, a deterrent. This is handy in some ways because the presence of caffeine, say in a water source, is a good sign that humans have been there. There is so much caffeine flushed through the pipes under the United States that it can be measured in the oceans on either side of it. Perhaps the fish are sleep-deprived as a result.

All the elaborate ways and means of getting caffeine into the bloodstream have developed attendant rituals, some of them beautiful. A Japanese tea ceremony can hardly be described simply as a means to access the drug 1,3,7-trimethylxanthine—a name you don't see much in lights on the menu board at Starbucks, but it's the real star of their show. There is a lot of money at stake in maintaining a mystique around coffee. Coffee is the second most valuable commodity traded in the world; only oil is more significant in the global economy. The difference is that the countries that produce oil tend to be rich. The countries producing coffee tend to be considerably poorer.

While the Western world likes to make a fuss about its coffee obsession, while there are prizes for making art in the milk froth on top of your coffee, there are real justice issues lurking in that delectable black drink. Consider this one scenario: over the last twenty years, Vietnam has come from nowhere to emerge as the second-largest coffee producer in the world, specializing in the cheaper and hardier Robusta varieties that tend to find their way into the tins and jars of instant coffee that, despite the efforts of connoisseurs, still account for far and away the lion's share of the market. After large parts of Vietnam were defoliated with Agent Orange during the Vietnam war, the Washington-based World Bank arranged for the place to be replanted. Sounds nice so far. But it was replanted with cheap coffee. The point was that this new source of cheap coffee was created to keep the traditional suppliers of coffee to the United States, mostly in Central and South America, on the ropes.

The international coffee business,
Brazil to Kenya, is murky. It d
producers who have not p

Antony Wild, auth
two tea producers
two tea consumers are
likely to be produced in t
a better chance of a reasonab
fee-producing nations are all in th
sumers, with the exception of Brazil,
Europe. The basic law of international tra
is able to pay a fair price, it doesn't mean the
Fair Trade Coffee movement is starting to make
teeing a price to producers. But next time you get a ch
how much shelf space in your supermarket is devoted to f
as opposed to the other types. And remember that when you op
instant coffee and get that satisfying aroma, a gas has been packed i
top of the jar just before it is sealed so that you'll get the whiff. It doe
come from the coffee.

Coffee excites rare passions. Kopi Luwak, a type of coffee produced in certain parts of Asia, including Sumatra and the Philippines, is highly prized among sophisticated drinkers. The beans of the coffee are gathered from the dung of the Asian palm civet, which eats those beans; the civet's digestive processes imitate the finest coffee preparation techniques, leaving the perfect bean vacuum-packed in a turd. In Vietnam it is called fox-dung coffee. But ask yourself who it was who first went poking around in civet poo on the off chance of a better cup of coffee. There you have a real coffee lover.

of coffee shares something of this strangeness. It's
ever made it to the cup in the first place. Raw coffee
n which the beans (or seeds) are found, are so unappe-
ethod of extracting the beans from the cherries, before
oasting and grinding and brewing them, is so involved that
how anyone thought of doing it. Coffee cherries are like
an't just pick them and use them. They need to be educated.
associates the origin of tea with Asia, of chocolate with South
, and of coffee with northern Africa. One popular story, if you
lieve it, concerns an Ethiopian goatherd called Kaldi who noticed
ock chewing on the coffee cherries and then getting frisky. So he
d them himself and felt pretty good as a result. He reported this to
e head of a monastery, who was wary of this devilish intoxicant and
ordered the bushes to be thrown on the fire. The delightful aroma of the burning bush—the world's first roasted beans—convinced him that coffee must be more divine than satanic, and before long, the monks had enlisted coffee in the service of God, using it to stay awake for prayers.

Whatever the story, there is a fair chance that coffee-drinking originated within Sufi communities near Harar in Ethiopia; the process of preparation appealed to the Sufi interest in alchemy. Sufism wanted to work the raw material of humanity into something divine, and drinking coffee began life as a form of communion with God; it was used in nocturnal rituals for much the same reason it is still used—that is, to apply a little pressure to nature in the hope of transcending it.

Like Arabic numerals, coffee (the "wine of Araby") is one of the many gifts of Islamic culture to the west. The emigration of coffee began from the Yemeni port of Mocha, and it had found its way to Constantinople, the fulcrum between east and west, by 1555. There's a legend that coffee was approved for Christian consumption by Clement VIII (pope from

1592 to 1605), who thought coffee was too good to be the devil's drink. He obviously hadn't tried the cheap stuff.

Coffee was unknown to Shakespeare at the time of his death in 1616. By 1621, Robert Burton—an an Oxford clergyman and one of the great magpies of intellectual history—described it as a curiosity in his omnivorous masterpiece, *The Anatomy of Melancholy*, a work that remains among literature's most strenuous attempts at self-understanding. Burton was prone to depression and thus was drawn to means to help "drive away the time." Burton defines sleep as "a rest or binding of the outward senses and of the common sense, for the preservation of the body and soul." Sleep is, for Burton, the time when "phantasy alone is free," and this was, for him, a frightening prospect. He speaks of coffee as a kind of rumor from the east, a possible respite from melancholy:

> The Turks have a drink called coffee . . . they spend much time in those coffee houses which are somewhat like our alehouses or taverns, and there they sit chatting and drinking to drive away the time, and to be merry together, because they find by experience that kind of drink, so used, helpeth digestion and procureth alacrity.

Another coffee legend arrived in 1626, when an early coffee adopter, Sir Thomas Herbert, traveled in Persia and returned with a story that the angel Gabriel had used coffee to keep the prophet Muhammad awake while he was dictating the Qur'an.

Within fifty years, coffee was no longer a novelty in Europe but a way of life. Writing in the *Spectator* in 1711, the essayist Richard Steele noted that "the Coffee-house is the Place of Rendezvous to all that live near it and who are thus turned to relish calm and ordinary Life." Ironically,

Steele saw coffee as a calming influence, enjoyed by men "in quiet Possession of the present Instant, as it passes, without desiring to quicken it by gratifying any Passion, or prosecuting any new Design."

In fact, caffeine led to many new designs. It changed history. The Boston Tea Party is a perfect example: an event that helped create the modern world by establishing an enduring demarcation whereby the English prefer tea and the Americans coffee. Numerous ideas, firms, deals, and revolutions were fermented (or perhaps brewed) in coffeehouses in Europe. Voltaire was part of the furniture of a café in Paris called Le Procope; in the same city, two hundred years later, Sartre was a fixture in the Café de Flore. There's a fair argument that, in a period when only a fool would consume the water, the entire Enlightenment owes its existence to the availability of something to drink other than alcohol. Coffee became an alternative to grog. As the name of the Enlightenment suggests, the world was suddenly awake.

These days, it is struggling to stay that way. Oceans of coffee have become part of the struggle. Ironically, people trying to become more alert during the day are often better off divorcing themselves from what may have become a dependent relationship. This is easier said than done; caffeine is a drug of addiction, which means it takes prisoners. Withdrawal symptoms include headaches the size of the Sahara. Caffeine hangs around in the body for a long time. But without it, you may sleep better and hence wake better.

When I saw an advertisement for a course to become a barista, I decided to give it a try. The instructor introduced himself to the eight of us taking the class, most of whom were young people looking for speedy access to part-time work, and announced that the single day of the course would acquaint us only with how much more we needed to learn: it would take a lifetime of practice and experience to be able to make a decent cup of coffee. "You need to give your life to master this art," he said solemnly.

I wasn't convinced, but we were clearly in the hands of an enthusiast. For him, espresso was "expressed," just like mother's milk.

"See this," he said, rubbing the side of the espresso machine. "You can have some beautiful moments with this machine. Beautiful moments."

It soon became clear that the course was not so much about how to make coffee as how to become better people, using coffee as a prop.

"The greatest day in the history of this city was May 1st, 1954," he said. "That was the day they opened Il Capuccino at 61c Fitzroy Street, St. Kilda. It was the day the first espresso was sold in this town."

In fact, coffee was well and truly available from overnight street vendors in Melbourne in the 1870s and '80s. By the turn of the century, coffee palaces were a significant part of the temperance movement, some of them built rather grandly to resemble the hotels to which they were designed to offer an alternative. But espresso was something different, and although early espresso machines may have done a sly trade from grocer's stores in Carlton, there is no doubt that 1954 is the key year in the coffee habits of the city. When a family called the Bancrofts imported the first espresso machine ever seen in Melbourne for Il Capuccino (spelled with one *p* in contrast to Il Cappuccino, the London establishment on which it was modeled), the event made news in the afternoon paper.

The instructor paused for a moment of reverence. "I am sorry that Il Capuccino has closed now."

Some of the participants in the course were starting to look agitated, a bit like people who've responded to a flyer for a free vegetarian meal and find themselves expected to join a cult. We were told to never, never buy coffee in the morning because it takes a machine half a day to properly warm up. This would seem to me to defeat the whole purpose of coffee, but here was a man who could be passionate about coffee for an entire day and never once mention caffeine. Nor did he refer to the circumstances under which most of the world's coffee is produced. But several times he reiterated the vital importance of serving latte in a glass, not a cup, and of tying a napkin around it to act as a handle.

"Coffee is something which engages all your five senses," he said. "If you want to make good coffee, you have to learn how to trust your senses. You have to listen to the coffee talk to you as it brews. You have to look at the color of the crema on top. And throw away those thermometers. If you want milk to froth at the perfect temperature, you have to learn to trust what your hand experiences as it holds the jug."

We learned about the fifty species of coffee, which fall largely into two types: Arabica and Robusta. Arabica is the better-quality coffee, grown at higher altitudes and providing a lower yield per acre. It is the coffee used in espresso. Robusta coffee comes from hardier plants that can be grown at a greater range of altitudes. Robusta varieties are the mainstay of instant coffee. It was clear we wouldn't be spending much time worrying about Robusta on this course. Each cup of coffee requires about six or seven grams of roasted beans.

Beans should be roasted between 392 and 464 degrees Fahrenheit and ground immediately before use. The grind of the coffee should be such that an espresso takes twenty-five seconds to work its way through the filter. Too fine and the extraction will be slow and the coffee will taste burnt; too coarse and the extraction will be fast and the coffee will lack body. A normal coffee shop could get through fifty-five pounds of coffee a week; a really busy one might need more than two hundred pounds.

I quickly did the math. That's between eight hundred and three thousand cups a day.

As much as the instructor despised Robusta coffee, he had a special contempt for machines that did all the work for you, including grinding the beans and frothing the milk, the type of machines you find in fast-food drive-thrus.

"Everyone needs to empower themselves," he explained. "You need to empower yourself to use your own senses. Have faith in your physical hand, your physical nose, your physical eyes. You must be the driver of the espresso machine. Don't let it drive you. You can be an artist or a slave—you choose. If the machine is making your decisions, you are its slave. In any café, the person standing behind the coffee machine is the one with real power. That is the command post of the entire business."

After lunch, we got practical and started making coffee. There was quite an agenda of styles we had to get through: latte, affogato, doppio, macchiato, mocha, ristretto, cappuccino, long black. Each student held their best effort aloft like a communion cup and was applauded by the group.

"You came to learn a skill," the instructor concluded. "I just hope that today you have been empowered."

6:45AM

[1978]

Some years ago, before I became a priest, I helped as a volunteer at The Way, a community for the homeless and needy in Fitzroy, a colorful inner suburb of Melbourne. One of our regulars was Johnny Foster. It was said that Johnny had been an officer in the army. He often slept out in a slouch hat and army coat, and he was a man who, despite his fair share of life's misfortunes, always carried himself with a military bearing.

There were countless stories about Johnny. One was that, in May 1978, the funeral of the late Sir Robert Menzies—Australia's longest-serving prime minister—had started ten minutes late because of Johnny Foster. The story went that the Prince of Wales had flown out overnight to read at the service. That same night, Johnny had been sleeping in one of the many haunts where, in the middle of a big city, people can make themselves invisible. Menzies was being buried from the Scots Church on the corner of Russell and Collins Streets, and a guard had been posted there to lend dignity to the occasion and to keep back the crowds. Just as the cavalcade bearing the Prince of Wales was nearing its destination, Johnny—stately in his great coat—stepped through the guard, took up position in the center of the intersection, and proceeded to direct the official party around the wrong corner. Johnny had such a commanding air that the drivers obeyed. So the cavalcade did another lap of the block, and Menzies was laid to rest a bit later than he may have been otherwise.

It was that coat that did the trick, making people think Johnny was official, and Johnny Foster sure was inseparable from his coat. He usually slept out and when he returned to The Way punctually at 6:45 AM, he would empty the contents of his coat pockets onto the kitchen table: he unearthed a whole host of found items, from Band-Aids to bricks. Other times, the coat pockets revealed more mysterious items: rosary beads, a police badge, even, on one confusing occasion, cutlery from the exclusive Melbourne Club. Johnny's coat was his caravan. He both traveled in it and slept in it. And where he'd disappear to at night was unknown by anyone. Where he slept was not our business. The street is a private address.

Just because you're homeless, it doesn't mean you don't have to sleep. Camping in the woods is one thing. But sleeping out in the city requires special skills. I once knew a man who slept between two shrubs on the median strip in the middle of the busiest road out of Melbourne. He believed that the key to becoming invisible was to make yourself as obvious as possible. Three lanes of traffic sped past in either direction but nobody ever noticed him; he slept here in all seasons for at least eighteen months, and then, suddenly one day, he vanished. He'd tried plenty of other things before he found this solution; he'd slept in forgotten corners of parks and hidden parts of the under-structure of bridges, but the more out of the way his camp, the sooner either the local thugs or police found him. In the plantation in the middle of a highway, he was safe. The only problem, he said, was getting women to have sex there. But then again, he conceded, he was finding it difficult to get women to have sex anywhere. Last time we spoke, he was considering a change of clothes to improve his chances.

Garbage dumpsters are another possibility for shelter in the night, as long as they are clean and don't get emptied during the night. I knew another man who slept in a cardboard and paper recycling bin behind a firm of architects. The cardboard was warm and certainly more comfortable than concrete, and he always had papers to read. The first person to arrive at the office every morning had the job of making him a cup of coffee and bringing it out. One year, he was invited to the office Christmas party but declined on the grounds he was a conscientious atheist who had existential objections to celebrating birthdays for gods that don't exist. The worker who'd made the invitation said that she needed a dictionary to figure out what he meant. The man happened to have a small one in his bag and lent it to her. Indeed, he traveled with English, Greek, and Spanish dictionaries. It was a pity he didn't go to the party, as the partners in the firm could have learned from him. He once said his experience of homelessness made him think a lot about architecture.

After an absence of some twenty years from Fitzroy, I found myself back in the same neck of the woods. The area had changed several times over in the meantime, but there is something stubborn about the House of Welcome, an establishment around the corner from the community of homeless people, where I used to live. The House of Welcome has stood by while the rest of the street has seen fashions come and go as cafes open, flourish for a while, fade, and then get made over, always in the search of the perfect cup of coffee. In all that time, the House of Welcome, a familiar place not far from The Way, hasn't changed its menu because it doesn't have a menu; the cuisine often includes food donated by local cafes. And coffee. Buckets of hot, cheap coffee with so much sugar you can stand a spoon in the mug. Every Monday morning, a few boys from the school where I teach come to help. They don aprons, put a small cup of cheap coffee powder into a huge pot, add water, and do the rounds. Breakfast is free to all patrons who show up.

The boys seldom drink coffee with the clients; they tend to nip into the Korean grocery next door where they can buy cheap Red Bull in cans covered with Chinese characters. The whole world drinks caffeine, but the way we get it divides us every which way.

At one level, the boys come to the House of Welcome to lend a hand. But really the purpose is for them to meet the people they most need to

At school, I am teaching George Orwell's *Nineteen Eighty-Four*, a work that begins with a clock striking thirteen. It's the period before lunch, and teenagers don't learn much on an empty stomach. At least it is better than the period after lunch, when they're prone to falling asleep—although to be frank, returning to the classroom after an interval of eighteen years, I was taken aback by seeing a few students dozing off at any time. I am too vain to sheet this all home to the dreariness of my teaching. There are other reasons. Some of the students have part-time jobs that keep them up late midweek, which is a sad state of affairs. Some of those need the money to pay for their phone and Internet usage, which keeps them up even later once they get home. The deputy head has discovered that students struggling at school are invariably spending two or three hours a night online, sometimes longer. This is not just exhausting. It also strikes me as lonely, spending so much time with friends who may or may not exist.

Orwell's book is a chilling study of a world deprived of language and memory. His vision of the future is in many ways an account of our present. Our public vocabulary is shrinking as so much conversation has become an exchange of clichés. The very use of the word *excellence*, for example, has become a sign of mediocrity. The culture we are part of is shifting from one based on memory to one based on *retention*. The difference is this. If you take twenty dollars out of an ATM, that little factoid will be retained for all eternity in the bowels of some machine. But if you then take the twenty dollars and buy some flowers for your beloved, a gesture of reconciliation or renewal or fondess or hope, the flutter in your heart can only be remembered by you and the person close enough to feel those exquisite little vibrations. It can't be *retained*. Memory is a human art, part of a relationship. Retention is a form of management. The devolution from a memory culture to a retention culture is an understandable way of coping with the sheer excess that is exhausting us all. But it is sad. The word *exhaustion* has its roots in the Latin *haustus*, which means a drawing of water and, by association, a drink. Exhaustion is the drought of the soul. It is a parched land.

The capacity for memory and language are both formed in sleep. All our lives, memory and language are consolidated in sleep. A culture that doesn't sleep properly uses more words to say less and has less capacity to remember. It may provide plenty of exhausting stimulation to distract us from our real needs. It may want us to live like bees in a bottle. But take a minute to look at a bee in the open air hovering over a flower. Observe the art of its stillness. This is when the bee is creative.

In *Nineteen Eighty-Four*, one character, O'Brien, a party member who seems for a time to offer a ray of hope to the depressed hero, Winston Smith, says to Smith, "We shall meet in the place where there is no darkness." Smith is uplifted by these words and sustained by them. But it turns out that the place where there is no dark is a vision of hell. It is the Ministry of Love, a vast windowless bunker operated by the Thought Police where torture and brutality take place. It is a place of despair and inhumanity. The lights are never turned out. Darkness would have been a mercy.

It concerns me that the young people I teach are growing up in a world of blinding light. They spend hours looking into little screens. Consider this difference. When you read something on a screen, such as on a computer, you are looking into a source of light. Light triggers various things in the brain, among them the creation of cortisol, the stress hormone, in the adrenal gland. On the other hand, if you read text in a book, you are looking at black ink. This has the opposite effect in the brain. Darkness facilitates the release of melatonin from the pineal gland. In some ways, melatonin does the opposite of cortisol: it is the hormone that urges calm, the one that regulates the circadian rhythm and helps the central nervous system not to go into overload. This is why reading a book in bed before going to sleep is a physically different experience from sitting up against the pillows with your laptop. Different hormones are called out to play. Cortisol is a good and necessary thing in its place. But we need breaks from it as well. Students who spend their lives looking at screens as opposed to books are learning in a more stressful environment. I have no idea why schools encourage this.

I tell my students that I prefer them to submit handwritten work. It's not that I want to make plagiarism more difficult. It's not even that, if they work on the computer, I have no idea if they can spell or not. It's really because I want them to have a closer relationship with their own work. I'd like to restrict access to the stress on which the world expects them to become addicted. They look at me like I've come from another planet.

BEDTIME

[2008]

"Benny, I think we need to finish the story now and go to sleep."

"No."

"Benny, it's time to put out the light."

"No. No."

"Benny. You've told us a beautiful story, but it's time for sleep now."

"But, Daddy!"

"No buts, Benny."

"But daddy, we're only up to the ending."

We're late to bed because it's a Friday night at the end of January and the summer holidays here in Australia are almost over and the peach tree out back is so heavy with fruit that the branches have bowed to the ground and we are wondering if the branches will snap before their load ripens. The passionfruit are ripe and the plums are perfect: the color of night on the outside and of the sunset on the inside. Benny (now four and a half) wants to lead a pirate expedition again; he calls himself Captain Orders, which means Jacob (two and a half) and Clare (still two minutes younger than Jacob) have to do what he says, although Clare wants to take off her clothes and climb the peach tree in the nude and Jacob has gone off to find his telescope, perhaps to check on what's up in the tree. There's no reason in the world to go inside except that the longer we put it off, the harder it gets. The students will be back at school on Monday after the long break, and I have to teach a novel I haven't read yet, so I want the kids in bed. But I also don't want this moment to end. I am wishing the time away at the same time as hoping it will last forever.

It's hard work to conjure sleep when the kids' imaginations are in flood. We manage to find room for them in the bath around a flotilla of pirate ships. Clare decides to give her Dorothy the Dinosaur a bath but then refuses to go to bed without her, so Dorothy has to be wrapped in three diapers so that she won't soak the bed and this means finding pink safety pins, as nothing else will be acceptable in Clare's pink princess bed. Two feet away, Jacob won't go to bed without a treasure map and one of his pirate books. He wants the little book but we can't find it, so he agrees to have the big one with pop-up pictures.

In the bunk on top of Jacob, Benny is worried. He's a concentrated version of his old man. I see my own anxiety in my little boy and worry about it. Tomorrow Benny is having a friend stay the night, the first time he has hosted a sleepover. It is a huge thing in his life. For the umpteenth time, we go through the arrangements. Clare will be coming into our room, which we have made to sound like a treat for her despite the fact she always ends up there at some time anyway. She will *start* the night tomorrow in our room in a special makeshift princess bed, every detail of which she has personally approved, including which side of the pillow will be facing upward, leaving us with the question of where she will go on her nocturnal wanderings when she hasn't got us as a destination. In the meantime, Benny has rung his friend Bobby to check what pajamas he will be bringing (Lightning McQueen) and to make sure he remembers his toothbrush (Buzz Lightyear) and swimming costume (which has a surfing dog on it). Bobby will be sleeping in Clare's bed and Benny needs reassurance once again that every trace of princess habitation will be expunged to be replaced by his old Buzz Lightyear pillow case, which won't match Bobby's pajamas but will at least match his toothbrush, which, if he wants to, Bobby can put on the window sill.

"How does a star fall softly?" asks Clare out of nowhere.

"That's a good question," replies Benny in an adult voice. "It means it lands on soft grass or soft weeds or a pillow someone has put out for it."

Clare is satisfied with this.

It's a hot night. Benny wants a fan. Jacob doesn't. Clare just wants to take off her clothes again and get back up in the peach tree. We whisper to Benny that we will bring in the fan when Jacob has fallen asleep. He likes this idea; it's a conspiracy. We read *Peter Pan's Snow Adventure*, a chastening tale of Captain Hook on ice. Then I tell a made-up story about Captain Hook, Cowboy Benny, and the Pink Princess in which Hook takes his ship up a long river to visit first a castle where he puts up a hook for the princess to hang her gown and then a wild west saloon where he puts up a hook for cowboy Benny to hang his hat when he goes to sleep, which, the storyteller prompts, should be right about now.

Mummy kisses good night. Daddy brings in his sitting chair from the kitchen. The lights go out.

Jacob starts crying.

Benny starts calling for Mummy. There is something else he needs to know about Bobby's sleeping arrangements for tomorrow.

Clare has tossed Dorothy out of the princess bed for being wet.

Now both boys are crying.

Clare rotates through 180 degrees, then back the other way through 90, then around again another 270. This is how she searches for sleep, turning and turning on the mattress like a ballet dancer until she nods off. The noise of the boys in the room ebbs and flows like waves on the beach; it builds to a crescendo then softens then comes rushing back over the sand. I try not to get cranky and tell them to be quiet because I'm going to write this in a book and I want them to read it later and think I was a gentle daddy who never lost his patience, least of all at ten o'clock when he has a novel to read for school. So I sit there, tense and rigid, as the noise bangs about my head.

Finally, the room is quiet, and I decide that reading the novel can wait. If need be, I can just put a flutter of yellow Post-it notes in it to show that I'm a world expert on the subject and go on the offensive, demanding to know why the students haven't read it over the holidays.

The Cistercian monk Thomas Merton wrote, "The night, O My Lord, is a time of freedom."

I remember when I was four and a half myself, trying to find sleep on hot nights. After she turned out the big light, Mum would spray the room with insecticide, which wafted like incense as my brother and I hid beneath the sheets, the last ritual of day. Mum said good night and muttered a blessing for us and a curse on all mosquitoes. When she was gone, I parted my pillow down the middle so my guardian angel could have half. Then I lay still and watched the shadows play over the ceiling until I summoned the courage to ask my angel if we could swap because her side of the bed was cooler. My guardian angel was always a girl, and she always obliged. I fell asleep listening to Mum and Dad say the rosary on the other side of the wall, starting in tired and cranky voices but gradually slowing to a gentle rhythm. Years later, I became superior to this kind of prayer and inwardly scoffed at it, but I have come around in my older age. I can see now that Mum and Dad's rosary was a kind of lovemaking and that they shared their intimate space with God. My kids have given me a second bite at innocence, and I owe them for that. I gave up on guardian angels as well for a while because I thought I was smart, but then we had three children of our own and I realized how much they need and accept friends of all kinds, angels included. For the freedom to believe without answers, I can thank the nights I have shared with little people and the big Whoever who found me there and reassured me even when I felt hopeless to the task of being a father. Merton says, "You, Who sleep in my breast, are not met with words."

zzzzzzz

A room filled with our three sleeping children is such a gift that I can't move. I mutter again the prayers that have already been answered as, little by little, the noises beyond the room come out of their hiding places. I can hear the fridge humming to itself in the kitchen and the washing machine clunk through another cycle as it deals with the clothes Clare threw out of the peach tree. Outside, a motorbike takes the neighbour off to work and the buses come back to roost in the depot opposite our gate. Gradually, the traffic on the freeway becomes audible and the train line four blocks away also comes back into earshot. Even farther away, noises from the container terminal of the port at the end of our street seep into the room; it sounds like the sugar boat is leaving the dock. We got to know our city better on the day we realized that, most weeks, a shipload of sugar weighs anchor at the bottom of the street next to ours. A horn announces that the town is ready for another sugar hit. There is always something more to know, so I try to stop thinking before thinking robs me of yet another present moment. Finally, I get to hear the noise from farther away than anything. I can hear my heart. It says that it's home. It says that it wants to sleep.

PAST BEDTIME

[2016]

Fatigue fatigue is when you're tired of being tired. Jenny and I have it bad. We promise each other an early night because the kids are in bed and it's been another long day and there are no more phone calls to make. But first there just might be something on television. At 9:00 PM, we find the remote where it has ended up with the dirty plates beside the sink and put the TV on. It's like sticking something in your arm. An hour later we are still there, too weary to make the effort to go to bed, too tired to sleep, unable to do anything more with the day but unwilling to let it go. There's a glass from dinner resting beside the screen where the remote usually sleeps. We are hungry for conversation but tiredness has robbed us of any appetite for it. But soon enough we will perform the nightly rituals: emptying the dishwasher, putting out clothes for tomorrow, brushing teeth, checking on the kids.

We lock the doors and the whole world scales down to the size of a house. As we finish in the kitchen and bathroom, the size of our world gradually diminishes again. We go to our room and it shrinks further. We get into bed and it is reduced again, now to a few square meters. We turn off the light and each of us is now reduced to dominion over half a bed. Finally, we surrender. It is an act of faith in the existence of tomorrow. We fall asleep. The moment that happens, the whole world mysteriously expands.

ON MY BEDSIDE TABLE
SOURCES FOR *SNOOZE*

Sleepfaring: A Journey through the Science of Sleep, Meir H.
Voman's Guide to Sleep Disorders, Meir H. Kryger, Thomas
illiam C. Dement's *Principles and Practice of Sleep Medi-*
artin's *Counting Sheep: The Science and Pleasures of Sleep and*
Carlos H. Schenck's *Sleep: The Mysteries, the Problems, and the*
special and poignant interest is Dien Dang, David Cun-
d John Swieca's article in the March/April 2011 edition
Neuropharmacology entitled "The emergence of devastating
ntrol disorders during dopamine agonist therapy of restless
me."

10:00PM, 1988

hanks to the boss who, in my first year of teaching, interrupted
ht preparations for class and told me simply, "Just remember:
id to talk to kids. You can't do that when you're tired." It was
ver forgot. Other resources for this chapter included Stanley
eep Thieves, Paul Kalanithi's *When Breath Becomes Air*, and Hugh
The Good Life.

10:30PM, 1980

)ement is a true pioneer in the field of sleep. His popular work
autobiographical. I highly recomend Dement and Christo-
ghan's *The Promise of Sleep: A Pioneer in Sleep Medicine Explores*
Connection Between Health, Happiness and a Good Night's Sleep. For
rested in more on William Blake, I recommend Peter Ackroyd's
sible biography entitled *Blake*.

8

Snooze is not a book of clinical
suggestions about sleep includ
ical School Guide to a Good Night
This Book Will Make You Sleep;
and Gerard T. Lombardo's *Slee*
Live Affects How Much You
Shows," by Inga Ting in *The Age*,

8:48

Thanks to the librarian at the Aus
high and low until he found the
written by Edison. Edison would
excellent resources include Neil
Mark Essig's Edison and the Electric C
tion; and *Robert Lomas's The Man W*

9:00PM

A great place to start a voyage with
Dead: Why Homer Matters. For the best
Fagles; Penguin, 2004) and *The Odyss*

9:45PM

Everyone is an expert on sleep scien
of your questions answered by any
Ackerman's *Sex Sleep Eat Drink Drea*

Jim Horne's
Kryger's *A*
Roth and W
cine; Paul M
Dreams; and
Solutions. O
nington, an
of *Clinical*
impulse co
legs syndro

Profound t
my late ni
you are pa
advice I n
Coren's *Sl*
Mackay's

William D
is partly
pher Vau
the Vital
those int
indespe

11:00 PM, 2005

Thanks to Malcolm Ramsay, my veterinarian friend, for conversations about sleep and animals. For an account of sleeping livestock, see A. Roger Ekirch's *At Day's Close: Night in Times Past.* For avian sleep, see Jennifer Ackerman's *The Genius of Birds.*

11:20 PM, 28 BC

There is great fun to be had with Charlotte Higgins's *Latin Love Lessons: Put a Little Ovid in Your Life.* You'll work a bit harder for your fun but will be well rewarded by Mary Beard's *Laughter in Ancient Rome.* The translations I consulted here include *The Aeneid* (tr. Robert Fagles; Penguin, 2007); *The Aeneid* (tr. David West); and H. Rushton Fairclough's *Virgil in Two Volumes.*

11:59 PM, 350 BC

For my work here, I consulted mainly Aristotle's *On the Soul, Parva Naturalia, On Breath* (tr. W. S. Hett); René Descartes's *Discourse on Method and the Meditations* (tr. F. E. Sutcliffe); and David Hume's *Principal Writings on Religion* (ed. J. C. A. Gaskin). Additional resources I suggest to help one become acquainted with the outline of philosophical ideas include Simon Critchley's *The Book of Dead Philosophers*; David Edmonds and John Eidinow's *Rousseau's Dog*; Nicholas Fearn's *Zeno and the Tortoise: How to Think Like a Philosopher*; Jostein Gaarder's *Sophie's World*; James Garvey and Jeremy Stangroom's *The Story of Philosophy: A History of Western Thought*; A. C. Grayling's *Descartes: The Life of René Descartes and its Place in His Times*; Christopher Phillips's *Socrates Café: A Fresh Taste of Philosophy.* Two overarching collections on general philosophy that are quite good are *Porcupines: A Philosophical Anthology* (edited by Graham Higgin) and

The Philosophers: Introducing Great Western Thinkers (edited by Ted Honderich).

MIDNIGHT, 1999

For additional reading on the subject of time, see David Ewing Duncan's *The Calendar*, A. Roger Ekirch's *At Day's Close: Night in Times Past*, and Alexander Waugh's *Time*. For information related to the topic of *Arabian Nights*, I used in particular *Tales from the Thousand and One Nights*, edited by N. J. Dawood, and Malcolm Lyons' *The Arabian Nights: Tales of 1001 Nights* (tr. Ursula Lyons).

12:02 AM, 1915 etc.

Of the many excellent sources on the topic of grief, C. S. Lewis—who wrote that "grief is like a bomber circling around and dropping its bombs"—has one of the best in his 1966 book *A Grief Observed*. Other sources related to Lewis and J. R. R. Tolkein include Humphrey Carpenter's *J. R. R. Tolkien: A Biography*, Colin Duriez's *J. R. R. Tolkien and C. S. Lewis: The Story of a Friendship*, and Philip Zaleski and Carol Zaleski's *The Fellowship: The Literary Lives of the Inklings*. Sources for research on fairytales included Bruno Bettleheim's *The Uses of Enchantment: The Meaning and Importance of Fairy Tales* and Maria Tatar's *The Annotated Brothers Grimm*. For my work on brain science and the relationship between war, PTSD and sleep, I consulted Norman Doidge's *The Brain that Changes Itself: Stories of Personal Triumph from the Frontiers of Brain Science*, Kevin Gournay's *Post-Traumatic Stress Disorder: Recovery After Accident and Disaster*, Ray Parkin's *Into the Smother*, James Prascevic's *Returned Soldier: My Battles—Timor, Iraq, Afghanistan, Depression, and Post Traumatic Stress Disorder*, Oliver Sacks's *Hallucinations*, Pattie Wright's *Ray Parkin's Odyssey*, and an excellent article by Nicole Hasham in the August 22, 2016 issue of *The Age* entitled "'Compliant, groggy state:' In Nauru's Ghost Camps, Refugees Sleep Away the Pain."

1:50 AM, 2000

Useful sources regarding Shakespeare include Stephen Greenblatt's *Will in the World: How Shakespeare Became Shakespeare* and James Shapiro's *1599: A Year in the Life of William Shakespeare*. Some of the information in this chapter about beds and nighttime customs came from Eileen Harris's *Going to Bed*; Greg Jenner's *A Million Years in a Day*, and Lawrence Wright's *Warm and Snug: The History of the Bed*.

2:00 AM, 1856

George Pickering's *Creative Malady* is over forty years old, but it is hard to beat for its understanding of illness as a lifestyle accessory. Also recommended is G. K. Chesteron's "On Lying in Bed" (from *Eight Essayists*, edited by A. S. Cairncross). Sources for information related to the life and times of Florence Nightingale included Mark Bostridge's *Florence Nightingale*; Gillian Gill's *Nightingales: The Extraordinary Upbringing and Curious Life of Miss Florence Nightingale*; Lytton Strachey's *Eminent Victorians*; Cecil Woodham-Smith's *Florence Nightingale*; and Florence Nightingale's own *Notes on Nursing*.

2:10 AM, 1728

In addition to my trusty old editions of *Gulliver's Travels* and *Robinson Crusoe*—both well-thumbed paperbacks with broken orange-colored spines from the Penguin English Library published in the 1970s—I also consulted for this chapter Leo Damrosch's *Jonathan Swift: His Life and His*

World; *Jonathan Swift: A Modest Proposal and Other Writings* (ed. Carole Fabricant); *Swift: Gullivers Travels and Selected Writings* (ed. John Hayward); Jonathan Sacks's *Not in God's Name: Confronting Religious Violence*; and Peter Steele's *Jonathan Swift: Preacher and Jester*. The editions of the Qu'ran referred to in this chapter and elsewhere are *The Qu'ran: A New Translation* by Tarif Khalidi, put out by Viking Books in 2008, and 1974's *The Koran*, translated by N. J. Dawood.

2:15 AM, 2007

For more on Freud, see Sigmund Freud's *The Interpretation of Dreams* (tr. A. A. Brill) and Anthony Storr's *Freud: A Very Short Introduction*. Other resources for this chapter include two books by Karen Armstrong: *Islam: A Short History* and *Muhammad: A Prophet for Our Time*.

2:35 AM, 2007

A 1999 work by Rosanna Vic put out by the Narcolepsy and Overwhelming Daytime Sleep Society of Australia entitled "Sleep Too Much or Too Little? What is your problem? NODSS Guide to Sleep Disorders" proved to be very helpful here. Also invaluable were *Experiences at the Edge of Consciousness*, edited by Anna Faherty, and Margueritte Jones Utley's *Narcolepsy: A Funny Disorder That's No Laughing Matter*.

3:15 AM, 2014

My favorite translation of *The Iliad*, Robert Fagles's work, which has a fantastic introduction by Bernard Knox, was used here. Also referenced are Desmond and Mpho Tutu's important *The Book of Forgiving: The Fourfold Path for Healing Ourselves and Our World* and Desmond Tutu's *No Future Without Forgiveness*.

3:30 AM, 1860

For work on famous sleepers and wakers throughout history, I consulted Eluned Summers-Bremner, *Insomnia: A Cultural History.* Additional information came from *A Benjamin Franklin Reader* (edited by Nathan G. Goodman) and Margaret Thatcher's *The Downing Street Years.* Dickens himself was a great help in understanding his sleep patterns, as seen in his *Selected Journalism* (ed. David Pascoe) and *Sketches by Boz* (put out by London's Mandarin Publishing in 1991). Also useful were Peter Ackroyd's *Dickens: Private Life and Public Passions;* Jane Smiley's *Charles Dickens: A Life;* and Claire Tomalin's *Charles Dickens: A Life; as well as* John Cosnett's "Charles Dickens and Sleep Disorders" in *The Dickensian* (ed. Malcolm Andrews; Winter, 1997).

4:30 AM, 2007

My favorite translation of *Don Quixote* is the 2005 Vintage Books edition translated by Edith Grossman. See also William Egginton's *The Man Who Invented Fiction: How Cervantes Ushered in the Modern World* and Ilan Stavans's *Quixote: The Novel and the World.* For my work on Peter Pan, I consulted mainly Andrew Birkin's *J. M. Barrie and the Lost Boys* and Lisa Chaney's *Hide-and-Seek with Angels: A Life of J. M. Barrie.*

6:00 AM, 1851

For my work on Balzac, in addition to Balzac's own *The Physiology of Marriage,* I consulted Andre Maurois's *Prometheus: The Life of Balzac* and Graham Robb's *Balzac: A Biography.* Also referenced: *Selections from the Tatler and the Spectator of Steele and Addison,* edited by Angus Ross. A few worthwhile resources on the topic of caffeine and coffee: Bonnie K. Bealer and Bennett Alan Weinberg's The *World of Caffeine: The Science and Culture of the World's Most Popular Drug,* Andrew Brown-May's *Espresso: Melbourne Coffee Stories,* and Anthony Wild's *Coffee: A Dark History.*

THIRTEEN O'CLOCK, 1984

Orwell had as many blind spots as anyone, but his famous 1946 essay "Politics and the English Language" is still a brilliant description of an exhausted culture, one in which tired people can only mouth clichés. Other sources for this chapter include Josef Pieper's *Lesiure: The Basis of Culture* and Bertrand Russell's *In Praise of Idleness and Other Essays.*

BEDTIME, 2008

Merton entered my life in a time of loneliness and taught me the meaning of solitude. For that I owe him more than thanks. Essential references on the topic of silence include Diarmuid MacCulloch's *Silence: A Christian History*, Thomas Merton's "Fire Watch, July 4, 1952" in *The Sign of Jonas*; and Henry David Thoreau's "Night and Moonlight" (*The Oxford Book of Essays*, edited by John Gross).

ACKNOWLEDGMENTS

Special thanks to Iris Blasi, Fran Bryson, Nikki Christer, Lucy Costa Michael Costigan, the Costigan family, Tony Flynn, Hugh Flynn, Clar Forster, Alisa Garrison, Jenny Gleeson, Ted Guinane, Jude Hallam, Michael Heyward, Martin Kelly, Daniel Lazar, Judith Lukin-Amundsen, Rod Morrison, Sabrina Plomitallo-González, Libby Roughhead, Stephen Russell, Coralie Scott, Peter Steele, Chris Straford, John Swieca , David Winter, Christopher Worsnop, Arnold Zable and the four wonderful people who are the stuff my dreams are made of: Jenny, Benedict, Jacob, and Clare.